Multilevel Groundwater Monitoring of Hydraulic Head and Temperature in the Eastern Snake River Plain Aquifer, Idaho National Laboratory, Idaho, 2009–10

By Brian V. Twining and Jason C. Fisher

Prepared in cooperation with the U.S. Department of Energy
DOE/ID–22221

Scientific Investigations Report 2012–5259

U.S. Department of the Interior
U.S. Geological Survey

U.S. Department of the Interior
KEN SALAZAR, Secretary

U.S. Geological Survey
Marcia K. McNutt, Director

U.S. Geological Survey, Reston, Virginia: 2012

For more information on the USGS—the Federal source for science about the Earth, its natural and living resources, natural hazards, and the environment, visit http://www.usgs.gov or call 1–888–ASK–USGS.

For an overview of USGS information products, including maps, imagery, and publications, visit http://www.usgs.gov/pubprod

To order this and other USGS information products, visit http://store.usgs.gov

Contents

Contents—Continued

Figures

Figures—Continued

Tables

Conversion Factors, Datums, and Abbreviations and Acronyms

Conversion Factors

Multiply	By	To obtain
Length		
inch (in.)	2.54	centimeter (cm)
foot (ft)	0.3048	meter (m)
mile (mi)	1.609	kilometer (km)
Area		
square mile (mi^2)	2.590	square kilometer (km^2)
Pressure		
pound per square inch (lb/in^2)	6.895	kilopascal (kPa)
Density		
pound per cubic foot (lb/ft^3)	16.02	kilogram per cubic meter (kg/m^3)
Hydraulic conductivity		
foot per day (ft/d)	0.3048	meter per day (m/d)
Hydraulic gradient		
foot per mile (ft/mi)	0.1894	meter per kilometer (m/km)
Transmissivity*		
foot squared per day (ft^2/ d)	0.09290	meter squared per day (m^2/d)

Temperature in degrees Celsius (°C) may be converted to degrees Fahrenheit (°F) as follows:

$$°F=(1.8×°C)+32.$$

*Transmissivity: The standard unit for transmissivity is cubic foot per day per square foot times foot of aquifer thickness [(ft3/d)/ft2]ft. In this report, the mathematically reduced form, foot squared per day (ft2/d), is used for convenience.

Datums

Vertical coordinate information is referenced to the National Geodetic Vertical Datum of 1929 (NGVD 29).

Horizontal coordinate information is referenced to the North American Datum of 1927 (NAD 27).

Altitude and hydraulic head, as used in this report, refer to distance above the vertical datum.

Conversion Factors, Datums, and Abbreviations and Acronyms—Continued

Abbreviations and Acronyms

Abbreviation or acronym	Definition
ATR	Advanced Test Reactor Complex
bls	below land surface
CFA	Central Facilities Area
ESRP	eastern Snake River Plain
head	hydraulic head
INL	Idaho National Laboratory
INTEC	Idaho Nuclear Technology and Engineering Center
MFC	Materials and Fuels Complex
MLMS	multilevel monitoring system
NRF	Naval Reactors Facility
PBF	Power Burst Facility
PCC	Pearson correlation coefficient
psi	pounds per square inch
psia	pounds per square inch absolute
RWMC	Radioactive Waste Management Complex
TAN	Test Area North
USGS	U.S. Geological Survey

Multilevel Groundwater Monitoring of Hydraulic Head and Temperature in the Eastern Snake River Plain Aquifer, Idaho National Laboratory, Idaho, 2009–10

By Brian V. Twining and Jason C. Fisher

Abstract

During 2009 and 2010, the U.S. Geological Survey's Idaho National Laboratory Project Office, in cooperation with the U.S. Department of Energy, collected quarterly, depth-discrete measurements of fluid pressure and temperature in nine boreholes located in the eastern Snake River Plain aquifer. Each borehole was instrumented with a multilevel monitoring system consisting of a series of valved measurement ports, packer bladders, casing segments, and couplers. Multilevel monitoring at the Idaho National Laboratory has been ongoing since 2006. This report summarizes data collected from three multilevel monitoring wells installed during 2009 and 2010 and presents updates to six multilevel monitoring wells. Hydraulic heads (heads) and groundwater temperatures were monitored from 9 multilevel monitoring wells, including 120 hydraulically isolated depth intervals from 448.0 to 1,377.6 feet below land surface.

Quarterly head and temperature profiles reveal unique patterns for vertical examination of the aquifer's complex basalt and sediment stratigraphy, proximity to aquifer recharge and discharge, and groundwater flow. These features contribute to some of the localized variability even though the general profile shape remained consistent over the period of record. Major inflections in the head profiles almost always coincided with low-permeability sediment layers and occasionally thick sequences of dense basalt. However, the presence of a sediment layer or dense basalt layer was insufficient for identifying the location of a major head change within a borehole without knowing the true areal extent and relative transmissivity of the lithologic unit. Temperature profiles for boreholes completed within the Big Lost Trough indicate linear conductive trends; whereas, temperature profiles for boreholes completed within the axial volcanic high indicate mostly convective heat transfer resulting from the vertical movement of groundwater. Additionally, temperature profiles provide evidence for stratification and mixing of water types along the southern boundary of the Idaho National Laboratory.

Vertical head and temperature change were quantified for each of the nine multilevel monitoring systems. The vertical head gradients were defined for the major inflections in the head profiles and were as high as 2.1 feet per foot. Low vertical head gradients indicated potential vertical connectivity and flow, and large gradient inflections indicated zones of relatively low vertical connectivity. Generally, zones that primarily are composed of fractured basalt displayed relatively small vertical head differences. Large head differences were attributed to poor vertical connectivity between fracture units because of sediment layering and/or dense basalt. Groundwater temperatures in all boreholes ranged from 10.2 to 16.3°C.

Normalized mean hydraulic head values were analyzed for all nine multilevel monitoring wells for the period of record (2007–10). The mean head values suggest a moderately positive correlation among all boreholes, which reflects regional fluctuations in water levels in response to seasonality. However, the temporal trend is slightly different when the location is considered; wells located along the southern boundary, within the axial volcanic high, show a strongly positive correlation.

Introduction

The Idaho National Laboratory (INL) was established in 1949 by the U.S. Atomic Energy Commission, which is now the U.S. Department of Energy, for the development of peacetime atomic-energy applications, nuclear safety research, defense programs, and advanced energy concepts. The INL covers an area of about 890 mi^2 and overlies the west-central part of the eastern Snake River Plain (ESRP) in southeastern Idaho (fig. 1). Over half a century of waste disposal at the INL has resulted in measurable concentrations of contaminants in the ESRP aquifer beneath the INL. Contaminants include several radiochemical, inorganic, and organic constituents (Mann and Beasley, 1994; Cecil and

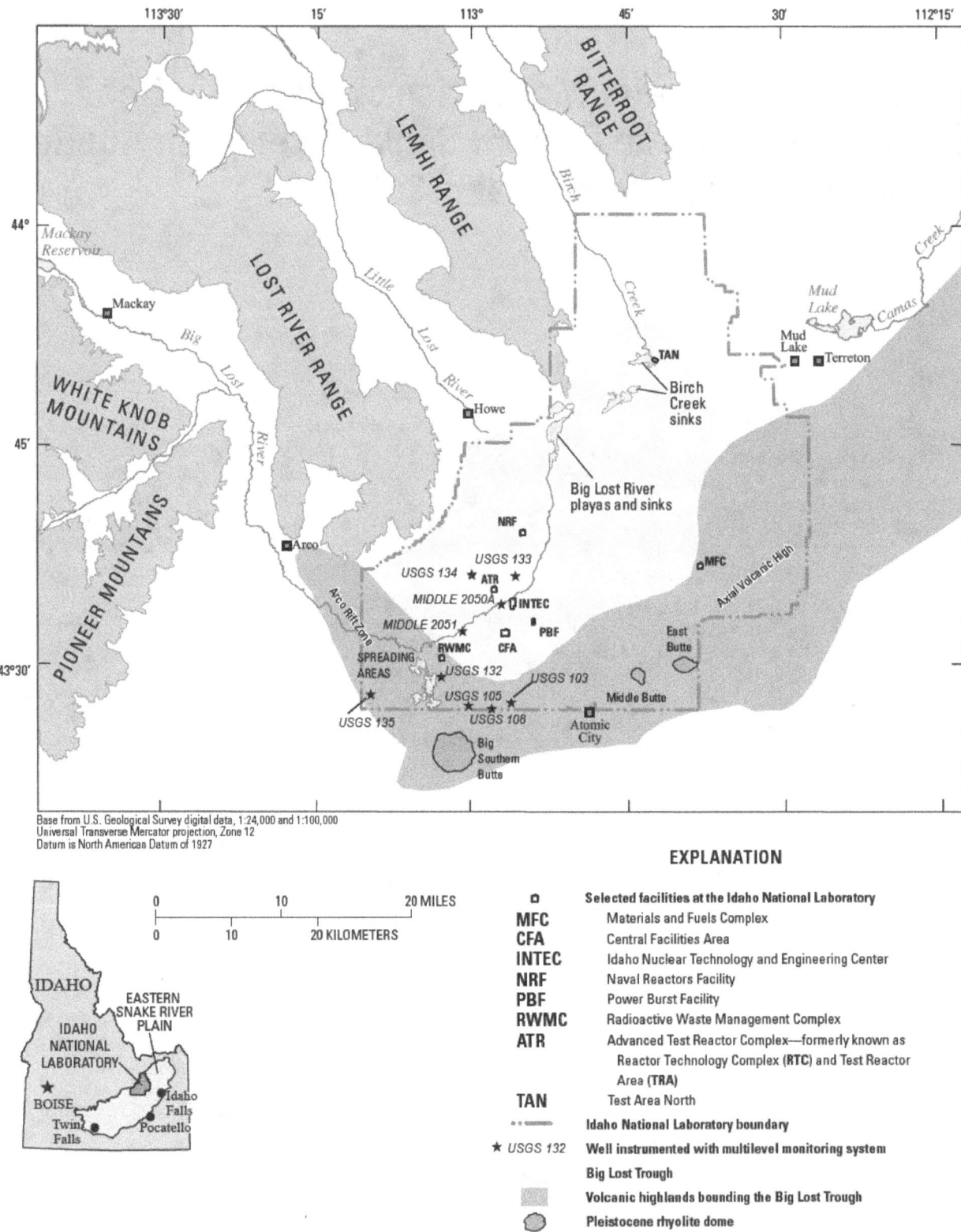

Figure 1.   Location of selected facilities, multilevel monitoring wells, and volcanic highlands bounding the Big Lost Trough, Idaho National Laboratory and vicinity, Idaho.

others, 1998; Bartholomay and others, 2000). The primary sources of contaminants are facility wastewater disposal sites, such as lined evaporation ponds, unlined infiltration ponds and ditches, drain fields, and injection wells. Determining the long-term risks associated with contaminants in the aquifer or that might be in the aquifer in the future is difficult because of slow releases of residual contamination in the unsaturated zone or waste buried in shallow pits and trenches.

Since 1949, the U.S. Geological Survey (USGS) has maintained a network of monitoring wells that record water levels and water quality in more than 200 boreholes with varying periods of record. Most monitoring wells are open boreholes, and groundwater flow is unrestricted into or out of the open wells (fig. 2). The fractured basalts of the ESRP aquifer are well-suited for this type of completion.

However, measurements collected from open-hole wells are independent of depth and represent a composite value that is a transmissivity-weighted average of all hydraulically conductive features in the borehole.

In 2005, the USGS INL Project Office began monitoring the vertical distribution of fluid pressures and chemistry using multilevel monitoring systems (MLMSs) completed within the ESRP aquifer. Previous monitoring information on six MLMSs was given in Fisher and Twining (2011). These data have been used to better characterize, manage, and remediate contaminated groundwater within the ESRP aquifer. The data analysis provides quarterly, depth-discrete measurements of vertical hydraulic head (head) and temperature in cored boreholes drilled to depths ranging from 818 to 1,427 feet below land surface (ft bls).

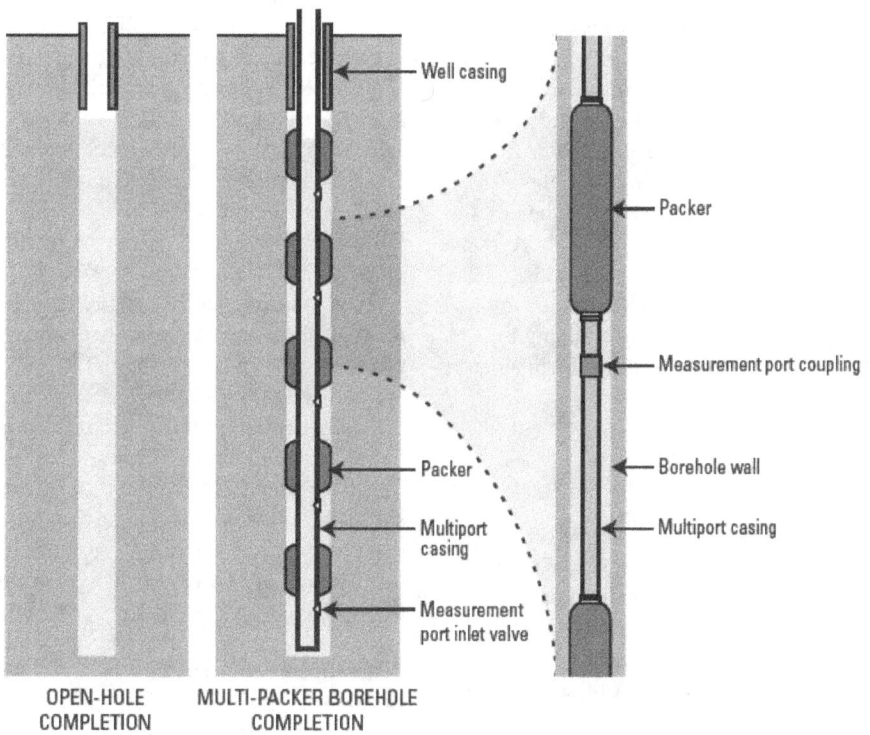

Figure 2.   Open-hole and multi-packer borehole completions, eastern Snake River Plain aquifer, Idaho National Laboratory and vicinity, Idaho.

Purpose and Scope

The purpose of this report is to summarize quarterly measurements of head and water- temperature data collected from nine MLMS boreholes during 2009 and 2010. This report also summarizes the methods used to collect depth-discrete measurements of hydraulic head and temperature from the nine MLMS boreholes along with a brief description of the lithology and multilevel completion design for three of the nine boreholes that were new completions during 2009–10. In addition, normalized mean hydraulic head values were analyzed for the period of record (2007–10) to conduct a comparative analysis of existing data.

The USGS INL Project Office collects various data from several hundred monitoring wells at the INL; however, most monitoring wells within the network are completed or screened into the ESRP aquifer to depths of less than 200 ft. The data collected from these wells are insufficient to accurately describe the vertical movement of water and contaminants in the ESRP aquifer, where downward contaminant movement persists. The MLMSs provide the necessary means to better characterize the areal extent and shape of contaminant plumes originating from the INL facilities because they allow for monitoring of head, temperature, and vertical chemistry concentrations in response to changes in geology. Completion depths for MLMS boreholes far exceed those of the average INL monitoring wells; therefore, any additional information pertaining to deeper flow and contaminant transport conditions will support ongoing numerical modeling efforts.

Geohydrologic Setting

The study area is in the ESRP in Idaho, a relatively flat topographic depression, about 200 mi long and 50–70 mi wide (fig. 1). The INL lies within the west-central part of the plain, and all MLMSs are within INL boundaries. Streams, some ephemeral, tributary to the ESRP and near the INL, originate in mountain ranges north and west of the study site and include the Big Lost River, the Little Lost River, Birch Creek, and Camas Creek. Streamflow-infiltration recharge fluctuates greatly in response to seasonality, such as spring snowmelt. Episodic recharge from the Big Lost River channel, spreading areas, sinks, and playas represent the largest transient stress within the ESRP aquifer at the INL (fig. 1). To prevent flooding of downstream facilities, a large percentage of the flow from the Big Lost River is diverted to a series of interconnected spreading basins near the southwestern

boundary of the INL (fig. 1). Episodic flood events can result in large pulses of surface-water infiltration near the southern boundary and have been shown to affect both the saturated and unsaturated zones in this region (Nimmo and others, 2002).

The ESRP is bounded by faults on the northwest and by downwarping and faulting on the southeast, and the basin has been filled with basaltic lava flows interbedded with terrestrial sediments. The basaltic rocks and sedimentary deposits combine to form the ESRP aquifer. Volcanic landforms of the ESRP include: (1) rhyolite domes (Kuntz and others, 1994), (2) sedimentary troughs (Gianniny and others, 1997), (3) vent corridors, and (4) volcanic highlands. The volcanic highlands are areas of focused volcanism resulting in high concentrations of volcanic vents and fissures (Anderson and others, 1999, p. 13; Hughes and others, 1999, p. 145), which are the major sources of basaltic rocks on the plain. A typical basalt flow has vesicular zones and cooling fractures on the top and sides, with vesicle sheets, pipe vesicles, mega vesicles in the interior, and a diktytaxitic to massive core (fig. 3). The Big Lost River has been the primary source of sediment since late Pliocene time, resulting in a depocenter known as the Big Lost Trough (fig. 1; Geslin and others, 2002). The Big Lost Trough contains significantly greater amounts of sediment than have been measured in boreholes in other parts of the INL (Anderson and others, 1999, fig. 9, table 2; Hughes and others, 2002; Welhan and others, 2007). Sediments penetrated by boreholes on the INL range in thickness from equal to or less than 1 to equal to or greater than 313 ft and are thickest in the northwestern part of the INL (Anderson and others, 1996; Welhan and others, 2007).

The ESRP aquifer is one of the most productive aquifers in the United States (U.S. Geological Survey, 1985, p. 193). The 2010 water-table contour(s), represented in figure 4, shows a southwestern regional flow direction in the aquifer that eventually discharges to springs along the Snake River downstream of Twin Falls, Idaho—about 100 mi southwest of the INL (fig. 1). Along the northwestern mountain front, surface-water and groundwater underflow enter the aquifer system from three tributary valleys—Big Lost River, Little Lost River, and Birch Creek. Groundwater moves horizontally through basalt interflow zones and contact between basalt flows, and vertically through joints and fracture zones (fig. 3). Infiltration of surface water, heavy pumpage, geologic conditions, and seasonal fluxes of recharge and discharge locally affect the movement of groundwater in the aquifer (Garabedian, 1986). Recharge primarily is from the infiltration of applied irrigation water, streamflow, precipitation, and underflow from the tributary valleys to the plain.

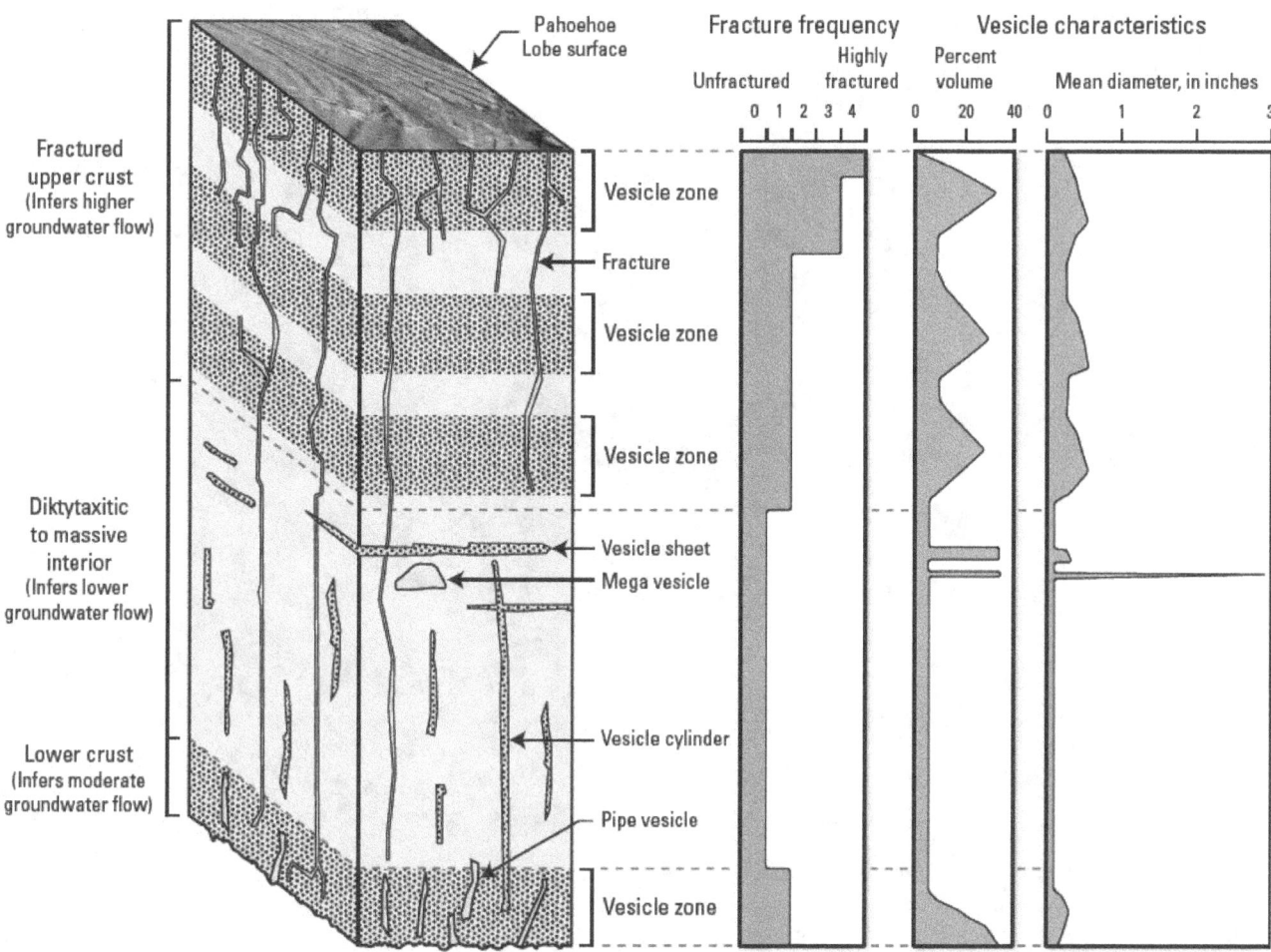

Figure 3.   Typical olivine tholeiitic pahoehoe basalt flow. (Modified from Self and others, 1998, p. 90, fig. 3.). The basalt flow is divided into three sections on the basis of vesicle characteristics and fracture frequency. Hydraulic conductivity is highest for the fractured upper crust, moderate for the lower crust, and lowest for the diktytaxitic to massive interior. Photograph of the pahoehoe lobe surface courtesy of Scott Hughes, Idaho State University, Pocatello, Idaho.

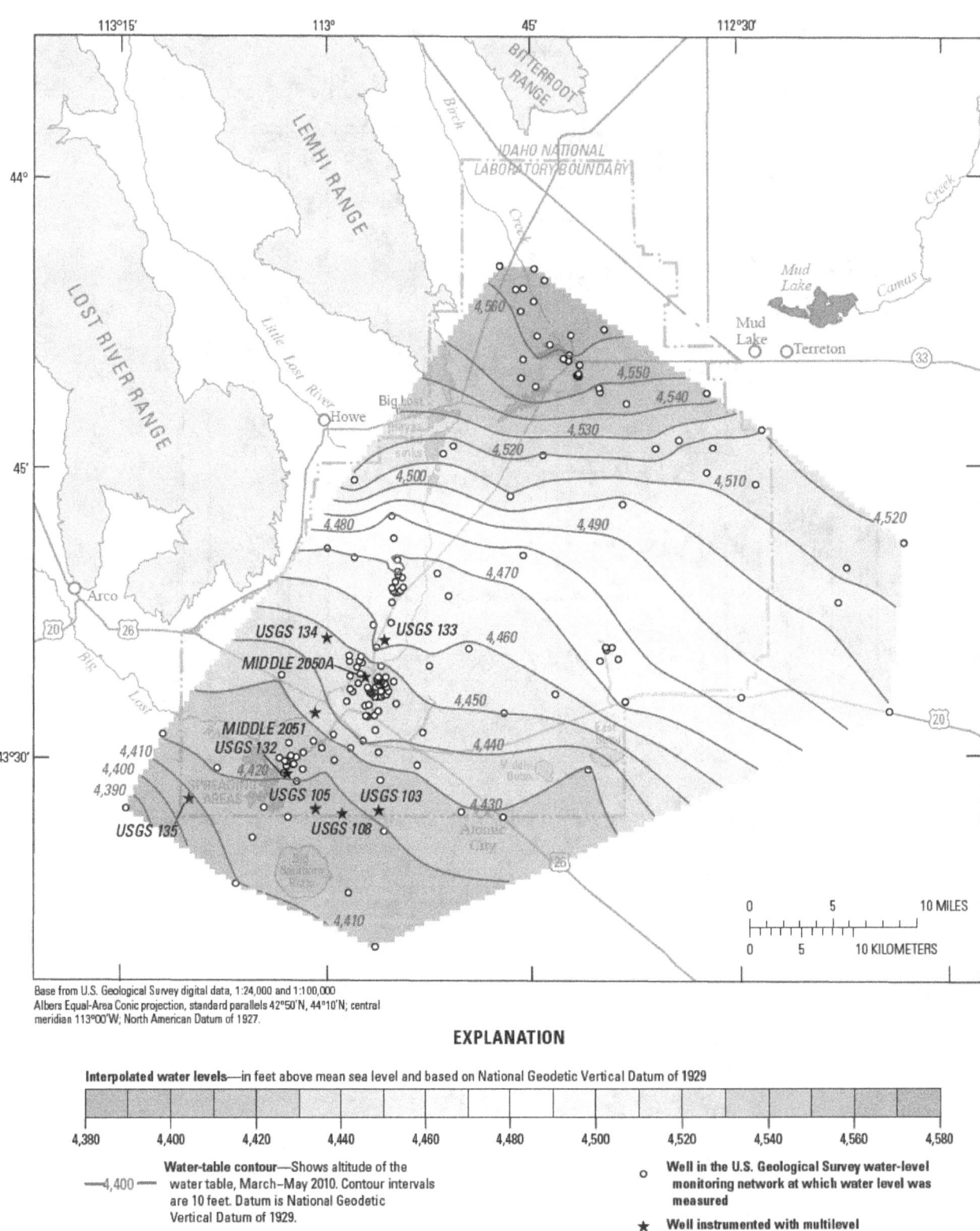

Base from U.S. Geological Survey digital data, 1:24,000 and 1:100,000
Albers Equal-Area Conic projection, standard parallels 42°50'N, 44°10'N; central
meridian 113°00'W; North American Datum of 1927.

EXPLANATION

Interpolated water levels—in feet above mean sea level and based on National Geodetic Vertical Datum of 1929

4,380   4,400   4,420   4,440   4,460   4,480   4,500   4,520   4,540   4,560   4,580

—4,400— **Water-table contour**—Shows altitude of the water table, March–May 2010. Contour intervals are 10 feet. Datum is National Geodetic Vertical Datum of 1929.

○   **Well in the U.S. Geological Survey water-level monitoring network at which water level was measured**

★   **Well instrumented with multilevel monitoring system**

Figure 4.   Water-table contours and monitoring wells, Idaho National Laboratory and vicinity, Idaho, March–May 2010.

Across the INL, borehole water-table altitudes range from about 4,560 to 4,410 ft (fig. 4). Depth to the water table ranges from about 200 ft BLS north of the INL to more than 900 ft BLS in the southeast. Ackerman (1991, p. 30) and Bartholomay and others (1997, table 3) reported a range of relative transmissivities for basalt in the upper part of the aquifer of 1.1–760,000 ft²/d. The hydraulic gradient at the INL generally flows from northeast to southwest and ranges from 2 to 10 ft/mi, with an average of about 4 ft/mi (Davis, 2010). Horizontal groundwater flow velocities ranging from 2 to 20 ft/d have been calculated based on the movement of various constituents in different areas of the aquifer beneath the INL (Robertson and others, 1974; Mann and Beasley, 1994; Cecil and others, 2000; Busenberg and others, 2001). Localized tracer tests at the INL have shown vertical and horizontal transport rates as high as 60–150 ft/d (Nimmo and others, 2002; Duke and others, 2007).

Previous Investigations

Several reports describing the geology and hydrology of the ESRP at the INL have been published; copies of these reports may be obtained from the USGS INL Project Office (U.S. Geological Survey, 2012). Water-quality data collected from MLMSs have been used to describe vertical movement of contaminants in the ESRP aquifer (Bartholomay and Twining, 2010), along with hydraulic head and temperature data (Fisher and Twining, 2011).

Fisher and Twining (2011) documented use of MLMSs to examine hydraulic head and temperature for six boreholes from 2007 to 2008. They described the MLMS components and specified the installation process. Additionally, they presented the methods used to construct head and temperature profiles and outlined quality-assurance methods that are summarized in this report.

Bartholomay and Twining (2010) documented the use of MLMSs to examine vertical changes in groundwater chemistry for six boreholes from 2007 to 2008. They examined water-quality samples from multiple water-bearing zones in the ESRP aquifer completed within about 350 –700 ft of the aquifer. The water-chemistry results were used to define movement of wastewater constituents in the aquifer.

Methods

The methods used to collect depth-discrete measurements of hydraulic head and temperature are described by Fisher and Twining (2011). Fisher and Twining (2011) also defined the modular system components (MP38 versus MP55), sampling probe, acquisition system, system dimensions, and installation of the MLMS. A general summary and update to the methods for this report include: (1) "Profiling and Completions," which describes the methods used to construct head and temperature profiles within a borehole; and (2) "Quality Assurance", which describes the accuracy and precision of head and temperature measurements.

Profiling and Completions

An individual head or temperature profile represents a set of measurements collected over a relatively short time period. The actual time required for each measurement period varied, and was dependent on the quantity and spacing of ports within a MLMS. Profile measurements in this study were less than 2 hours, a period considered instantaneous when contrasted to the slow response times of groundwater systems.

Fluid pressure and temperature measurements were made using a portable sampling probe, a wireline-operated probe that is lowered into the multiport casing from the land surface and positioned at a selected measurement port coupling (fig. 5). The positioned probe is then coupled with the measurement-port inlet valve to allow monitoring of groundwater outside the multiport casing and within the monitoring zone, so that groundwater in this zone is vertically isolated between upper and lower packers. Coupling the probe with the measurement port inlet valve is done by extending the backing shoe on the probe to create a hydraulic seal between the probe and the port and to open the port. Fluid pressure and temperature measurements are then transmitted to the land surface through the wireline communication cable, processed using a data acquisition system, and recorded on a datalogger. The hydraulic head at each measurement port, assuming 100 percent barometric efficiency, was expressed as:

$$H = \Psi_2 + Z - D = \left(\frac{P_2 - P_{Atm}}{\gamma_w} \right) \times 144 + Z - D \tag{1}$$

where

H is the hydraulic head, in ft,

Ψ_2 is the pressure head outside the multiport casing, in ft,

Z is the altitude of a referenced land – surface measurement point, in ft,

D is the depth to the pressure transducer sensor at the measurement port coupling, in ft bls,

P_2 is the fluid pressure measured outside the multiport casing in pounds per square inch absolute (psia),

P_{Atm} is the atmospheric pressure measured at land surface, in psia, and

γ_w is the specific weight of water, in lb / ft³.

Atmospheric pressure was monitored at the land surface using a hand-held barometric sensor. The specific weight of water was calculated as a function of temperature only (McCutcheon and others, 1993), assuming negligible salinity and gravitational differences between measurements, and expressed as:

$$\gamma_w = 62.42796 \times \left\{ 1 - \left(\left[\frac{T + 288.9414}{508929.2 \times (T + 68.12963)} \right] \times (T - 3.9863)^2 \right) \right\} \quad (2)$$

where γ_w is in units of lb/ft^3 and T is water temperature measured inside the multiport casing from the bridge of the pressure transducer in degrees Celsius. The depth to the pressure transducer sensor at a port coupling was measured once (appendix A) and calculated as:

$$D = \Psi_1 + L_1 = \left(\frac{P_1 - P_{Atm}}{\gamma_w} \right) \times 144 + L_1 \quad (3)$$

where

    D is the depth to the pressure transducer, in ft bls,

    Ψ_1 is the pressure head inside the multiport casing, in ft,

    L_1 is the depth to water inside the multiport casing, in ft bls, and

    P_1 is the fluid pressure measured inside the multiport casing, in psia.

The depth to water inside the multiport casing (L_1) was measured using an electronic measuring tape and corrected for borehole deviation. Simultaneous measurements of P_1, P_{Atm}, and L_1 were made at each port coupling to account for (1) temporal changes in atmospheric conditions, and (2) depth to water that was dependent on the volume of water displaced by the wireline communication cable.

Multilevel completions included the location of measurement port valves, port couplings, packers, and monitoring zones in the borehole (where a monitoring zone describes the volumetric space between consecutive packers outside the multiport casing). The location of a multilevel component in a borehole is based on the measured depth to the pressure transducer at a port coupling (eq. 3) and its position within the MLMS installation log. For example, the length of a monitoring zone is defined as the distance between two consecutive packer seals and calculated by subtracting the depth at the bottom of the upper packer from the depth at the top of the lower packer, or:

$$M_z = (D_{z-1} - a - b - c - d) - (D_z - a - b - c) = D_{z-1} - d - D_z \quad (4)$$

where

    M_z is the distance between packer seals in monitoring zone z, in ft,

    D_z is the depth to the pressure transducer sensor in the upper port coupling of zone z, in ft bls,

    D_{z-1} is the depth to the pressure transducer sensor in the uppermost port coupling of zone z – 1, the zone located directly beneath zone z, in ft bls;

    a is the distance between the pressure transducer sensor and the center of the measurement port inlet valve, in ft, 0.17 ft in both MP systems,

    b is the distance between the center of the measurement port inlet valve and the top of the measurement port coupling, in ft, 0.50 ft in the MP55 system and 0.38 ft in the MP38 system,

    c is the distance between the uppermost measurement port coupling and the bottom of the adjacent packer, in ft, 0.60 ft in both MP systems, and

    d is the thickness of the inflated packer seal, in ft, 3.00 ft in both MP systems.

Parameters a, b, c, and d were defined using nominal component lengths specified in the MLMS installation log (fig. 5); however, actual parameter lengths may vary because of component deformation in the multiport casing and port couplings that results from mechanical stretch and thermal expansion during MLMS installation. However, destressing during packer inflation was used to reduce the total strain on the system. Measurement errors associated with component deformation were assumed to be negligible because nominal component lengths were relatively small when compared to the measured depth to a pressure transducer (D).

Standard procedures for collecting profile measurements were first described in Fisher and Twining (2011). The steps are summarized as follows: (1) the sampling probe is lowered to the deepest measurement port in the MLMS; (2) the probe is coupled with the monitoring port to continuously monitor fluid pressure and temperature; (3) measurements of fluid pressure, atmospheric pressure, and water temperature are recorded on a field sheet (appendix B) after temperature readings stabilize with fluctuations of less than 0.1°C (generally in 30 minutes or less); and (4) after fluid pressure and temperature measurements are recorded, the probe is decoupled from the port and raised to the next highest measurement port. The process is repeated until all ports are measured and final measurements are recorded.

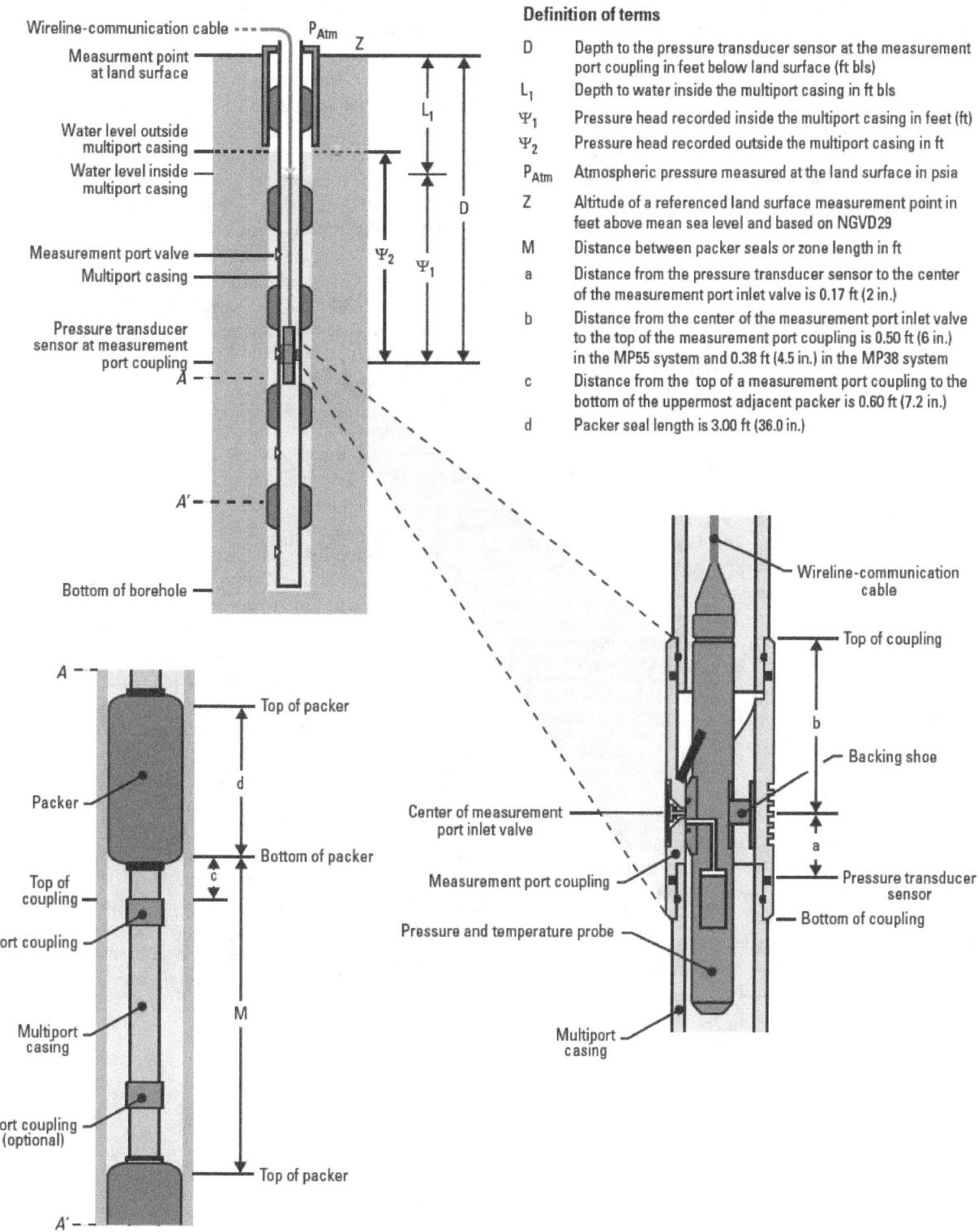

Definition of terms

D   Depth to the pressure transducer sensor at the measurement port coupling in feet below land surface (ft bls)

L_1 Depth to water inside the multiport casing in ft bls

Ψ_1 Pressure head recorded inside the multiport casing in feet (ft)

Ψ_2 Pressure head recorded outside the multiport casing in ft

P_{Atm} Atmospheric pressure measured at the land surface in psia

Z   Altitude of a referenced land surface measurement point in feet above mean sea level and based on NGVD29

M   Distance between packer seals or zone length in ft

a   Distance from the pressure transducer sensor to the center of the measurement port inlet valve is 0.17 ft (2 in.)

b   Distance from the center of the measurement port inlet valve to the top of the measurement port coupling is 0.50 ft (6 in.) in the MP55 system and 0.38 ft (4.5 in.) in the MP38 system

c   Distance from the top of a measurement port coupling to the bottom of the uppermost adjacent packer is 0.60 ft (7.2 in.)

d   Packer seal length is 3.00 ft (36.0 in.)

Figure 5. Terms used in the calculation of hydraulic head based on the portable probe position when coupled with a measurement port in the multilevel monitoring system (Fisher and Twining, 2011).

Custom computer programs were developed using the R programming language (R Development Core Team, 2010) to process and graph the head and temperature profiles in each borehole (Fisher and Twining, 2011). In theory, head is equivalent throughout a monitoring zone with flow dominated by the most transmissive feature penetrated by the borehole in this zone. Head values reported at monitoring zones containing a second measurement port reflect an average of the two head values.

Quality Assurance

The first effort to quantify the accuracy and precision of MLMS head measurements was described by Fisher and Twining, (2011). Hydraulic head was examined for five variables based on equations 1 and 2: (1) the water temperature outside the multiport casing measured indirectly by the temperature sensor inside the multiport casing and in the sampling probe; (2) the fluid pressure outside the multiport casing measured by the sampling probe; (3) the atmospheric or barometric pressure measured at the top of the multiport casing at the land surface; (4) the altitude of a reference point at the land surface; and (5) the vertical depth, or distance between the land surface reference point and the pressure transducer sensor in a measurement port coupling.

The cumulative error of all five variables for independent head readings is ±2.3 ft; a value determined by summing measurement accuracies for fluid pressure head (±1.15 ft), atmospheric pressure head (±0.01 ft), land-surface altitude (±0.01 ft), and pressure transducer sensor depth (±1.17 ft) (Fisher and Twining, 2011). Many of the sources of measurement error are diminished when considering the differences between two closely spaced readings of head, where head values are monitored using the same pressure probe, at similar depths, and at similar water densities. Under these conditions, vertical head differences have much less error than the error associated with any single head measurement because some sources of error subtract (for example, drift, offset, temperature effect) and are equal or nearly equal for adjacent port readings. Therefore, a ±0.1-ft measurement accuracy was assumed for vertical head differences (and gradients) calculated between adjacent monitoring zones.

Calibration of the fluid pressure sensor was performed by the probe manufacturer; calibration test results are shown in appendix C. Each test was run over a referenced pressure range from 15 to 500 pounds per square inch absolute (psia), with probe temperature held constant at about 10 and 20°C. Accounting for the range of fluid pressures measured in the field, from 30 to 350 psia, the calibration tests gave a standard deviation for the measurement accuracy of ±0.101 pounds per square inch (psi) (or ±0.23 ft at 10°C) and ±0.098 psi (or ±0.23 ft at 20°C). Tests indicate that fluid pressure error remained well below the specified accuracy of the sensor during the duration of the study. Calibration corrections were not applied to fluid pressure measurements because of the relatively high specified accuracy of the reference pressure sensor at ±0.100 psi (or ±0.23 ft at 13°C, which is the average ambient temperature in these boreholes).

The precision of the fluid pressure measurement was determined by comparing fluid pressure measurements between consecutive profiles. Repeat measurements were made for each of the nine MLMS boreholes, with two repeat measurements for each profile. A 0.01-ft mean difference between consecutive measurements and a 0.04-ft standard deviation indicate consistently high precision for the instrument. Measurement precision was tested again by comparing head values between paired ports, with two measurement ports located in the same monitoring zone (fig. 6). Theoretically, the distribution of head within a monitoring zone should be uniform; therefore, any significant head difference between paired-port measurements provides evidence of a malfunctioning measurement port, a well construction anomaly, or groundwater density variations because of differences in total dissolved solids. Paired–port head differences generally were small with an average value of 0.03 ft and a standard deviation of 0.17 ft. Relatively large head differences ranging from 0.28 to 0.81 ft were measured between paired–ports in monitoring zone 15 of borehole USGS 134. These head differences were first described in Fisher and Twining (2011) and were attributed to an improper seal while coupling the probe with port 20 and (or) a water density distribution within zone 15 that varied over space and time.

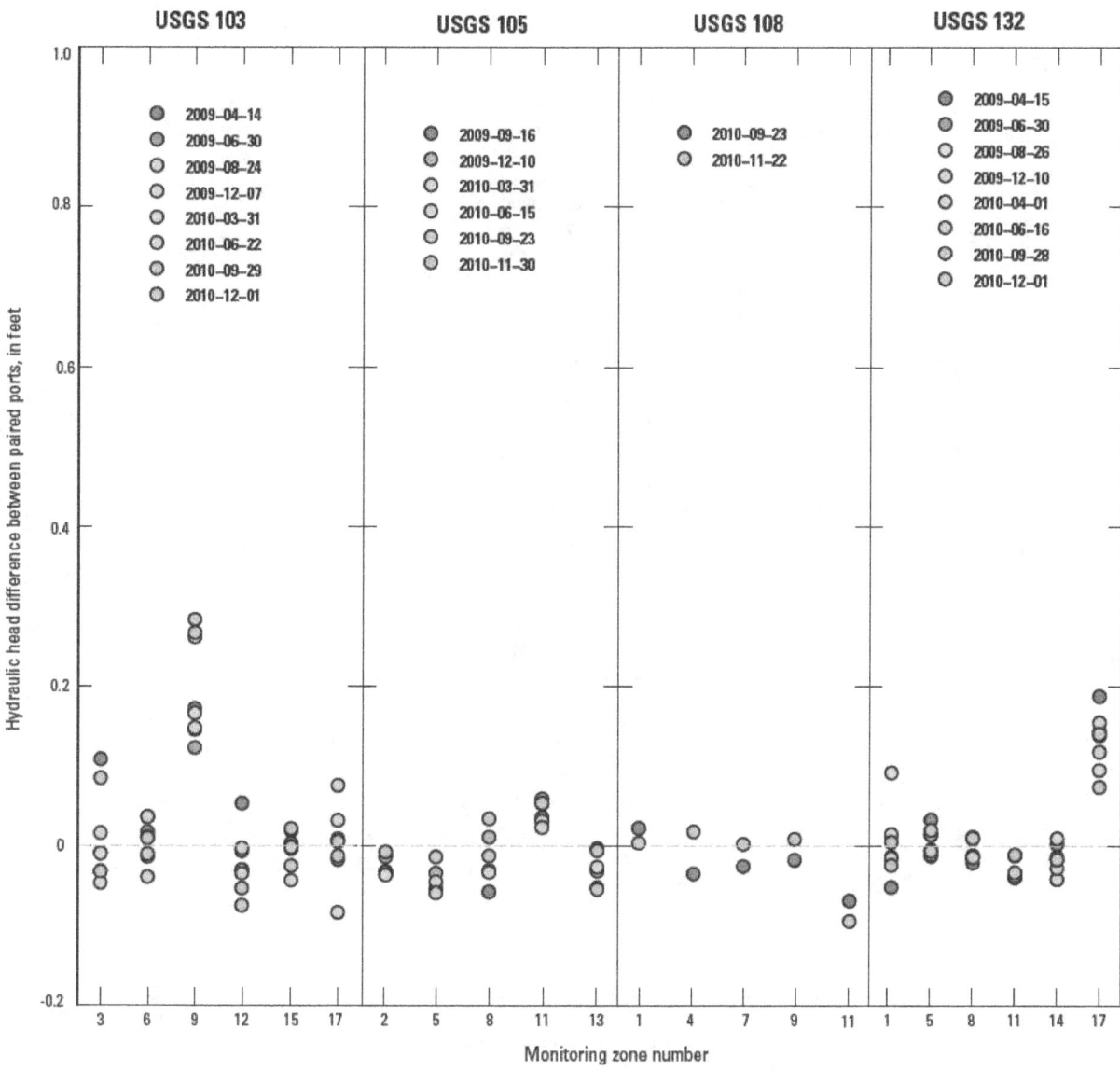

Figure 6. Hydraulic head differences between paired-ports and measurement ports in the same monitoring zone in boreholes USGS 103, USGS 105, USGS 108, USGS 132, USGS 133, USGS 134, and USGS 135, Idaho National Laboratory, Idaho, 2009–10.

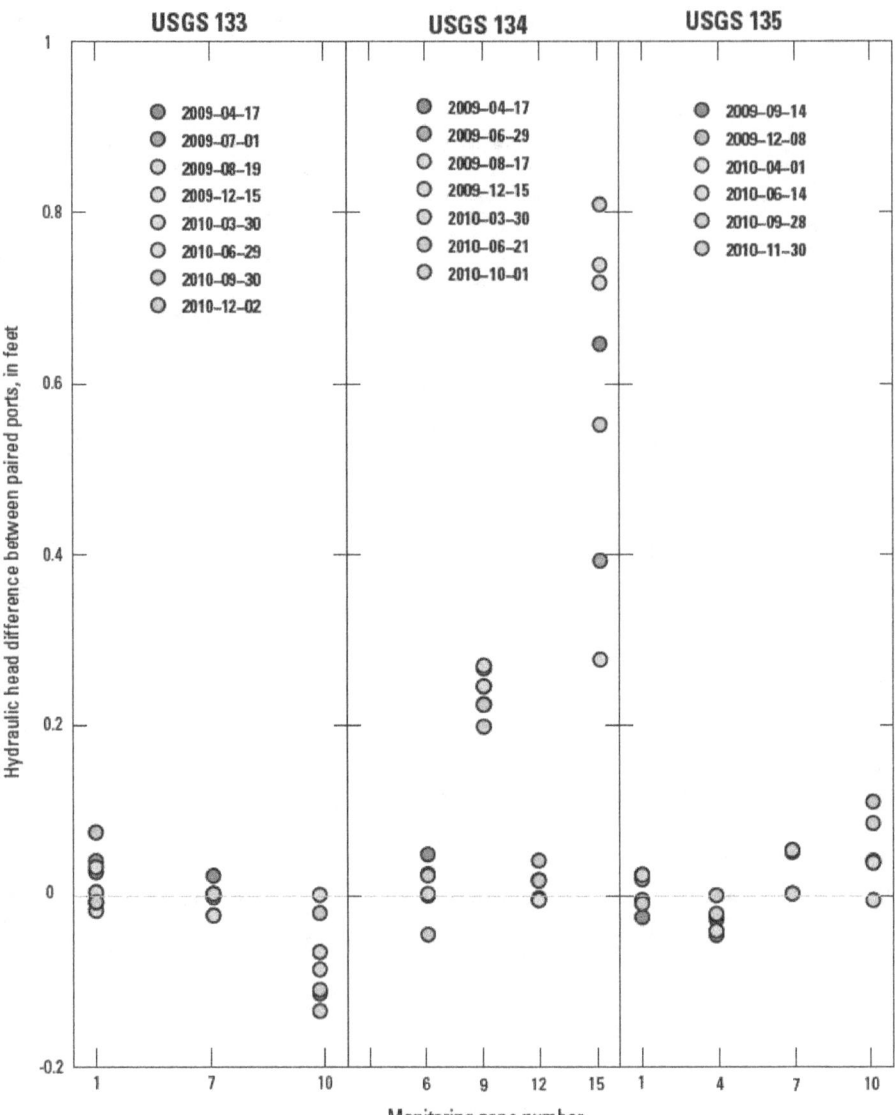

Figure 6.—Continued.

Hydraulic Head and Temperature Measurements

Hydraulic head and groundwater temperature measurements in the network of MLMSs are presented for 2009 and 2010. The nine boreholes instrumented with MLMSs include USGS 103, USGS 105, USGS 108, USGS 132, USGS 133, USGS 134, USGS 135, MIDDLE 2050A, and MIDDLE 2051 (fig. 1; table 1). Head and temperature measurements were recorded in 120 hydraulically isolated monitoring zones located 448.0–1,377.6 ft bls (table 2). Detailed descriptions of the geophysical traces, lithology log, completion log, and profiles are provided for boreholes USGS 105, 108, and 135. Geophysical descriptions for boreholes USGS 103, USGS 132, USGS 133, USGS 134, MIDDLE 2050A, and MIDDLE 2051 are provided by Fisher and Twining (2011). Profile shapes and inflection points were analyzed both temporally and spatially for each borehole.

Quarterly Measurements

During the 2009–10 multilevel monitoring period, 120 profiles were collected; these profiles represent 2,104 individual measurements of head and temperature from 9 MLMS boreholes (fig. 7; appendix D). With the exception of boreholes USGS 105, USGS 108, USGS 134, and USGS 135, profiles were measured each quarter during the 2009 and 2010 calendar years. Profiles were started in USGS 105 and USGS 135 after installation in the third quarter of 2009, and they were started in USGS 108 after installation in the third quarter

of 2010. No data were collected from USGS 134 in the fourth quarter of 2010 because weather conditions prevented access. Throughout the 2-year monitoring period, head at all MLMS boreholes ranged from 4,416.8 to 4,463.6 ft (appendix D), with the smallest head at USGS 135 and the largest head at USGS 133, respectively. The lowest head values were measured in the farthest downgradient boreholes USGS 103, USGS 105, USGS 108, USGS 132, and USGS 135 (near the southern boundary); the highest head values were in the farthest upgradient borehole USGS 133 (fig. 4). Water temperature ranged from 10.4 to 16.3°C at boreholes MIDDLE 2051 and MIDDLE 2050A, respectively (appendix D), which is within the reported range for temperatures measured in the ESRP aquifer at or near the INL—8.3–19.5°C (Davis, 2008).

In order to quantify the amount of temporal variability in MLMS head and temperature profile shapes for each borehole during 2009 and 2010, a Pearson correlation coefficient (PCC) was computed for each profile, as described by Fisher and Twining (2011). The PCC ranges from -1 to 1; however, the closer the PCC is to either -1 or 1, the stronger the correlation. To evaluate the correlation among all head or temperature profiles in a well requires the calculation of PCC for all permutations of profiles taken two at a time for each port; the minimum of these values reflects the poorest correlation between profiles in a borehole. The minimum PCC values associated with head and temperature profiles for each borehole are summarized in table 3. The aquifer's complex basalt and sediment stratigraphy, proximity to aquifer recharge and discharge, and groundwater flow contribute to some localized variability even though the general profile shape remained consistent over the measured time frame (fig. 7).

Table 1. Data for multilevel groundwater monitoring wells and boreholes, Idaho National Laboratory, 2009–10.

[**Local name** is the local well identifier used in this study. Location of boreholes is shown in figure 1. **Site No.** is the unique numerical identifiers used to access well data (http://waterdata.usgs.gov/nwis). **Latitude and Longitude** is in degrees, minutes, seconds and based on the North American Datum of 1927 (NAD27). **Land-surface altitude** is in feet above National Geodetic Vertical Datum of 1929 (NGVD29). **Base of aquifer altitude** is in feet above NGVD29 (Whitehead, 1992; Anderson and Liszewski, 1997). **Hole depth** is in feet below land surface (ft bls)]

| | | Boreholes | | | | |
Local name	Site No.	Latitude	Longitude	Land-surface altitude (ft)	Estimated base of aquifer altitude (ft)	Hole depth (ft bls)
USGS 103	432714112560701	43°27'13.57"	112°56'06.53"	5,007.42	2,470	1,307
USGS 105	432703113001801	43°27'03.40"	113°00'17.78"	5,095.12	2,540	1,409
USGS 108	432659112582601	43°26'58.79"	112°58'26.34"	5,031.36	2,495	1,218
USGS 132	432906113025001	43°29'06.68"	113°02'50.93"	5,028.60	2,540	1,238
USGS 133	433605112554301	43°36'05.50"	112°55'43.80"	4,890.12	3,960	818
USGS 134	433611112595801	43°36'11.15"	112°59'58.27"	4,968.84	3,960	949
USGS 135	432753113093601	43°27'53.47"	113°09'35.62"	5,135.94	2,675	1,198
MIDDLE 2050A	433409112570501	43°34'09.48"	112°57'05.38"	4,928.22	3,790	1,427
MIDDLE 2051	433217113004901	43°32'16.93"	113°00'49.38"	4,997.31	3,270	1,179

Table 2.   Data for multilevel well completions, boreholes USGS 103, USGS 105, USGS 108, USGS 132, USGS 133, USGS 134, USGS 135, MIDDLE 2050A, and Middle 2051, Idaho National Laboratory, Idaho, 2009–10.

[**Local name** is the local well identifier used in this study. Location of boreholes is shown in figure 1. **Site No.** is the unique numerical identifiers used to access port data (http://waterdata.usgs.gov/nwis). **Zone No.** is the identifier used to locate monitoring zones. **Zone depth interval** limits are in feet below land surface (ft bls) and length is in feet (ft). **Port No.** is the identifier used to locate port couplings. **Port coupling depth** is the depth to the top of the measurment port coupling in ft bls]

			Boreholes				
			Zone depth interval				**Port coupling**
Local name	**Site No.**	**Zone No.**	**Bottom (ft bls)**	**Top (ft bls)**	**Length (ft)**	**Port No.**	**depth (ft bls)**
USGS 103	432714112560702	1	1,279.4	1,257.4	22.0	1	1,258.0
	432714112560703	2	1,254.4	1,242.9	11.5	2	1,243.5
	432714112560704	3	1,239.9	1,184.4	55.6	3	1,209.7
	432714112560705					4	1,185.0
	432714112560706	4	1,181.4	1,115.2	66.2	5	1,115.8
	432714112560707	5	1,112.2	1,100.6	11.5	6	1,101.2
	432714112560708	6	1,097.6	1,063.2	34.5	7	1,086.8
	432714112560709					8	1,063.8
	432714112560710	7	1,060.2	1,045.5	14.7	9	1,046.1
	432714112560711	8	1,042.5	1,016.5	26.0	10	1,017.1
	432714112560712	9	1,013.5	958.0	55.4	11	992.9
	432714112560713					12	958.6
	432714112560714	10	955.0	948.4	6.6	13	949.0
	432714112560715	11	945.4	922.6	22.8	14	923.2
	432714112560716	12	919.6	891.6	28.0	15	908.7
	432714112560717					16	892.2
	432714112560718	13	888.6	862.6	26.0	17	863.2
	432714112560719	14	859.6	835.1	24.5	18	835.7
	432714112560720	15	832.1	766.9	65.2	19	801.9
	432714112560721					20	767.5
	432714112560722	16	763.9	694.3	69.7	21	694.9
	432714112560723	17	691.3	669.6	21.7	22	680.3
	432714112560724					23	670.2
USGS 105	432703113001802	1	1,290.1	1,279.2	10.9	1	1,279.8
	432703113001803	2	1,276.2	1,224.8	51.4	2	1,242.2
	432703113001804					3	1,225.4
	432703113001805	3	1,221.8	1,165.9	55.9	4	1,166.5
	432703113001806	4	1,162.9	1,105.4	57.5	5	1,106.0
	432703113001807	5	1,102.4	1,034.6	67.8	6	1,071.6
	432703113001808					7	1,035.2
	432703113001809	6	1,031.6	1,005.1	26.5	8	1,005.7
	432703113001810	7	1,002.1	985.4	16.7	9	986.0
	432703113001811	8	982.4	929.3	53.1	10	951.6
	432703113001812					11	929.9
	432703113001813	9	926.3	909.6	16.7	12	910.2
	432703113001814	10	906.6	865.3	41.3	13	865.9
	432703113001815	11	862.3	830.4	31.9	14	851.2
	432703113001816					15	831.0
	432703113001817	12	827.4	754.9	72.5	16	755.5
	432703113001818	13	751.9	706.9	45.1	17	727.6
	432703113001819					18	707.4

Table 2. Data for multilevel well completions, boreholes USGS 103, USGS 105, USGS 108, USGS 132, USGS 133, USGS 134, USGS 135, MIDDLE 2050A, and Middle 2051, Idaho National Laboratory, Idaho, 2009–10.—Continued

[**Local name** is the local well identifier used in this study. Location of boreholes is shown in figure 1. **Site No.** is the unique numerical identifiers used to access port data (http://waterdata.usgs.gov/nwis). **Zone No.** is the identifier used to locate monitoring zones. **Zone depth interval** limits are in feet below land surface (ft bls) and length is in feet (ft). **Port No.** is the identifier used to locate port couplings. **Port coupling depth** is the depth to the top of the measurment port coupling in ft bls]

| Local name | Site No. | Zone No. | Zone depth interval | | | Port No. | Port coupling depth (ft bls) |
			Bottom (ft bls)	Top (ft bls)	Length (ft)		
USGS 108	432659112582602	1	1,191.9	1,160.9	31.0	1	1171.8
	432659112582603					2	1161.5
	432659112582604	2	1,157.9	1,121.6	36.3	3	1122.2
	432659112582605	3	1,118.6	1,062.6	56.0	4	1063.2
	432659112582606	4	1,059.6	1,018.0	41.6	5	1028.8
	432659112582607					6	1018.6
	432659112582608	5	1,015.0	980.4	34.6	7	981.0
	432659112582609	6	977.4	906.7	70.7	8	907.3
	432659112582610	7	903.7	872.0	31.7	9	887.7
	432659112582611					10	872.6
	432659112582612	8	869.0	832.7	36.3	11	833.3
	432659112582613	9	829.7	791.4	38.3	12	808.8
	432659112582614					13	792.0
	432659112582615	10	788.4	681.8	106.6	14	682.4
	432659112582616	11	678.8	642.1	36.7	15	661.1
	432659112582617					16	642.7
USGS 132	432906113025001	1	1,213.6	1,152.3	61.3	1	1,173.0
	432906113025002					2	1,152.9
	432906113025003	2	1,149.3	1,144.1	5.2	3	1,144.7
	432906113025004	3	1,141.1	1,134.3	6.8	4	1,134.9
	432906113025005	4	1,131.3	1,046.1	85.3	5	1,046.7
	432906113025006	5	1,043.1	984.3	58.7	6	1,011.6
	432906113025007					7	984.9
	432906113025008	6	981.3	953.2	28.2	8	953.8
	432906113025009	7	950.2	938.4	11.8	9	939.0
	432906113025010	8	935.4	911.1	24.3	10	918.7
	432906113025011					11	911.7
	432906113025012	9	908.1	876.7	31.4	12	877.3
	432906113025013	10	873.7	866.8	6.8	13	867.4
	432906113025014	11	863.8	811.5	52.3	14	827.3
	432906113025015					15	812.1
	432906113025016	12	808.5	801.6	6.9	16	802.2
	432906113025017	13	798.6	790.1	8.5	17	790.7
	432906113025018	14	787.1	726.6	60.5	18	765.4
	432906113025019					19	727.2
	432906113025020	15	723.6	672.5	51.1	20	673.1
	432906113025021	16	669.5	662.6	6.9	21	663.2
	432906113025022	17	659.6	623.6	36.1	22	637.9
	432906113025023					23	624.2

Table 2. Data for multilevel well completions, boreholes USGS 103, USGS 105, USGS 108, USGS 132, USGS 133, USGS 134, USGS 135, MIDDLE 2050A, and Middle 2051, Idaho National Laboratory, Idaho, 2009–10.—Continued

[**Local name** is the local well identifier used in this study. Location of boreholes is shown in figure 1. **Site No.** is the unique numerical identifiers used to access port data (http://waterdata.usgs.gov/nwis). **Zone No.** is the identifier used to locate monitoring zones. **Zone depth interval** limits are in feet below land surface (ft bls) and length is in feet (ft). **Port No.** is the identifier used to locate port couplings. **Port coupling depth** is the depth to the top of the measurment port coupling in ft bls]

| | | | Boreholes | | | | |
| | | | Zone depth interval | | | | Port coupling |
Local name	Site No.	Zone No.	Bottom (ft bls)	Top (ft bls)	Length (ft)	Port No.	depth (ft bls)
USGS 133	433605112554301	1	766.4	724.8	41.6	1	745.5
	433605112554302					2	725.4
	433605112554303	2	721.8	715.0	6.8	3	715.6
	433605112554304	3	712.0	698.6	13.4	4	699.2
	433605112554305	4	695.6	685.5	10.1	5	686.1
	433605112554306	5	682.5	618.2	64.3	6	618.8
	433605112554307	6	615.2	593.7	21.6	7	594.3
	433605112554308	7	590.7	555.5	35.2	8	569.6
	433605112554309					9	556.1
	433605112554310	8	552.5	539.1	13.4	10	539.7
	433605112554311	9	536.1	483.2	52.8	11	483.8
	433605112554312	10	480.2	448.0	32.3	12	469.1
	433605112554313					13	448.6
USGS 134	433611112595801	1	886.8	881.0	5.8	1	881.6
	433611112595802	2	878.0	871.0	7.0	2	871.6
	433611112595803	3	868.0	846.0	22.0	3	856.1
	433611112595804					4	846.6
	433611112595805	4	843.0	831.0	12.0	5	831.6
	433611112595806	5	828.0	821.0	7.0	6	821.6
	433611112595807	6	818.0	782.0	36.0	7	806.6
	433611112595808					8	782.6
	433611112595809	7	779.0	747.0	32.0	9	747.6
	433611112595810	8	744.0	723.0	21.0	10	723.6
	433611112595811	9	720.0	690.9	29.0	11	706.5
	433611112595812					12	691.5
	433611112595813	10	687.9	664.9	23.0	13	665.5
	433611112595814	11	661.9	654.9	7.0	14	655.5
	433611112595815	12	651.9	638.9	13.0	15	645.5
	433611112595816					16	639.5
	433611112595817	13	635.9	604.8	31.0	17	605.4
	433611112595818	14	601.8	592.8	9.0	18	593.4
	433611112595819	15	589.8	553.8	36.0	19	578.5
	433611112595820					20	554.4

Table 2. Data for multilevel well completions, boreholes USGS 103, USGS 105, USGS 108, USGS 132, USGS 133, USGS 134, USGS 135, MIDDLE 2050A, and Middle 2051, Idaho National Laboratory, Idaho, 2009–10.—Continued

[**Local name** is the local well identifier used in this study. Location of boreholes is shown in figure 1. **Site No.** is the unique numerical identifiers used to access port data (http://waterdata.usgs.gov/nwis). **Zone No.** is the identifier used to locate monitoring zones. **Zone depth interval** limits are in feet below land surface (ft bls) and length is in feet (ft). **Port No.** is the identifier used to locate port couplings. **Port coupling depth** is the depth to the top of the measurment port coupling in ft bls]

			Boreholes				
			Zone depth interval				Port coupling
Local name	Site No.	Zone No.	Bottom (ft bls)	Top (ft bls)	Length (ft)	Port No.	depth (ft bls)
USGS 135	432753113093601	1	1,136.6	1,105.6	31.0	1	1,116.4
	432753113093602					2	1,106.1
	432753113093603	2	1,102.6	1,054.8	47.8	3	1,055.3
	432753113093604	3	1,051.8	1,010.6	41.2	4	1,011.1
	432753113093605	4	1,007.6	967.5	40.1	5	988.1
	432753113093606					6	968.0
	432753113093607	5	964.5	923.3	41.3	7	923.7
	432753113093608	6	920.3	864.2	56.1	8	864.7
	432753113093609	7	861.2	822.6	38.6	9	836.7
	432753113093610					10	823.1
	432753113093611	8	819.6	790.0	29.7	11	790.4
	432753113093612	9	787.0	765.3	21.6	12	765.8
	432753113093613	10	762.3	727.0	35.3	13	737.9
	432753113093614					14	727.5
MIDDLE 2050A	433409112570501	1	1,377.6	1,267.5	110.1	1	1,268.1
	433409112570502	2	1,264.5	1,229.7	34.7	2	1,230.3
	433409112570503	3	1,226.7	1,179.7	47.0	3	1,180.3
	433409112570504	4	1,176.7	1,081.3	95.4	4	1,081.9
	433409112570505	5	1,078.3	1,043.6	34.7	5	1,044.2
	433409112570506	6	1,040.6	998.7	41.9	6	999.3
	433409112570507	7	995.7	843.1	152.6	7	843.7
	433409112570508	8	840.1	810.4	29.8	8	811.0
	433409112570509	9	807.4	790.0	17.4	9	790.6
	433409112570510	10	787.0	719.5	67.5	10	720.1
	433409112570511	11	716.5	706.4	10.2	11	707.0
	433409112570512	12	703.4	643.3	60.1	12	643.9
	433409112570513	13	640.3	623.7	16.6	13	624.3
	433409112570514	14	620.7	541.6	79.0	14	542.2
	433409112570515	15	538.6	464.9	73.7	15	516.8
MIDDLE 2051	433217113004901	1	1,176.5	1,140.3	36.2	1	1,140.9
	433217113004902	2	1,137.3	1,130.5	6.8	2	1,131.1
	433217113004903	3	1,127.5	1,090.5	37.0	3	1,091.1
	433217113004904	4	1,087.5	1,002.2	85.3	4	1,002.8
	433217113004905	5	999.2	879.4	119.8	5	880.0
	433217113004906	6	876.4	826.2	50.1	6	826.8
	433217113004907	7	823.2	791.9	31.4	7	792.5
	433217113004908	8	788.9	773.8	15.0	8	774.4
	433217113004909	9	770.8	748.4	22.4	9	749.0
	433217113004910	10	745.4	646.7	98.8	10	647.3
	433217113004911	11	643.7	612.2	31.5	11	612.8
	433217113004912	12	609.2	561.8	47.4	12	602.9

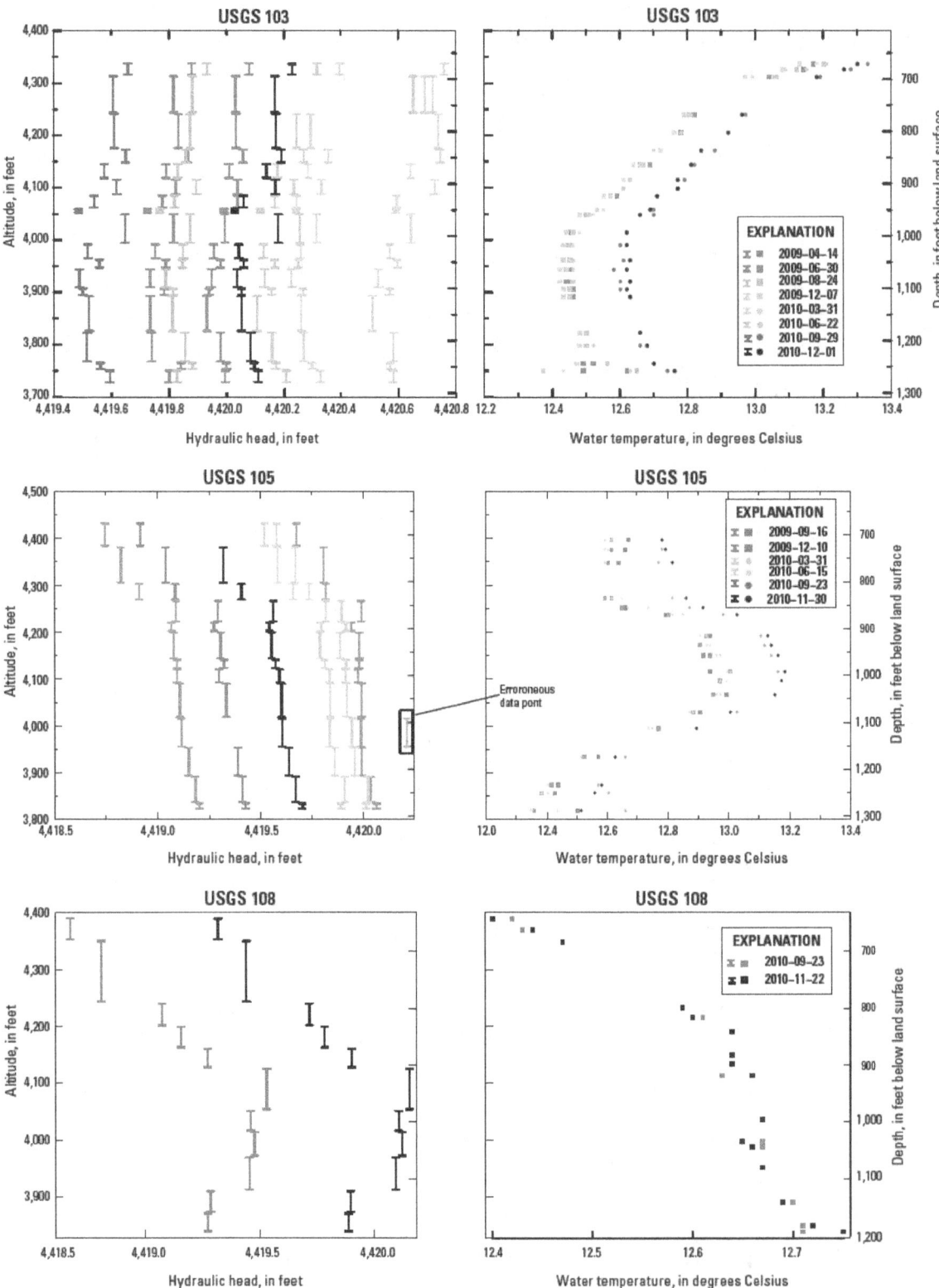

Figure 7.   Vertical hydraulic head and water temperature profiles at boreholes USGS 103, USGS 105, USGS 108, USGS 132, USGS 133, USGS 134, USGS 135, MIDDLE 2050A, and MIDDLE 2051, Idaho National Laboratory, Idaho. Profiles are based on quarterly measurements made during 2009–10.

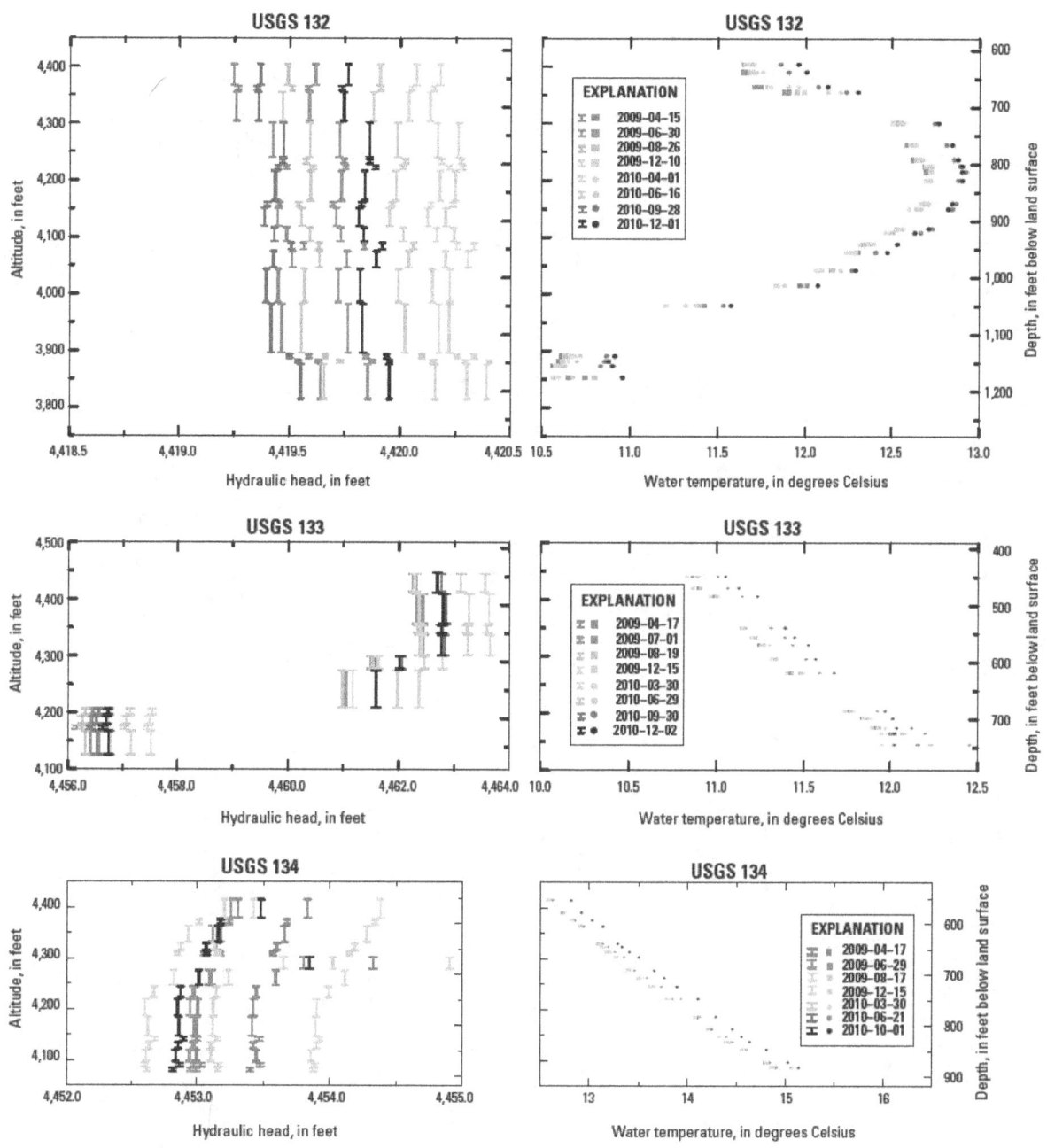

Figure 7.—Continued.

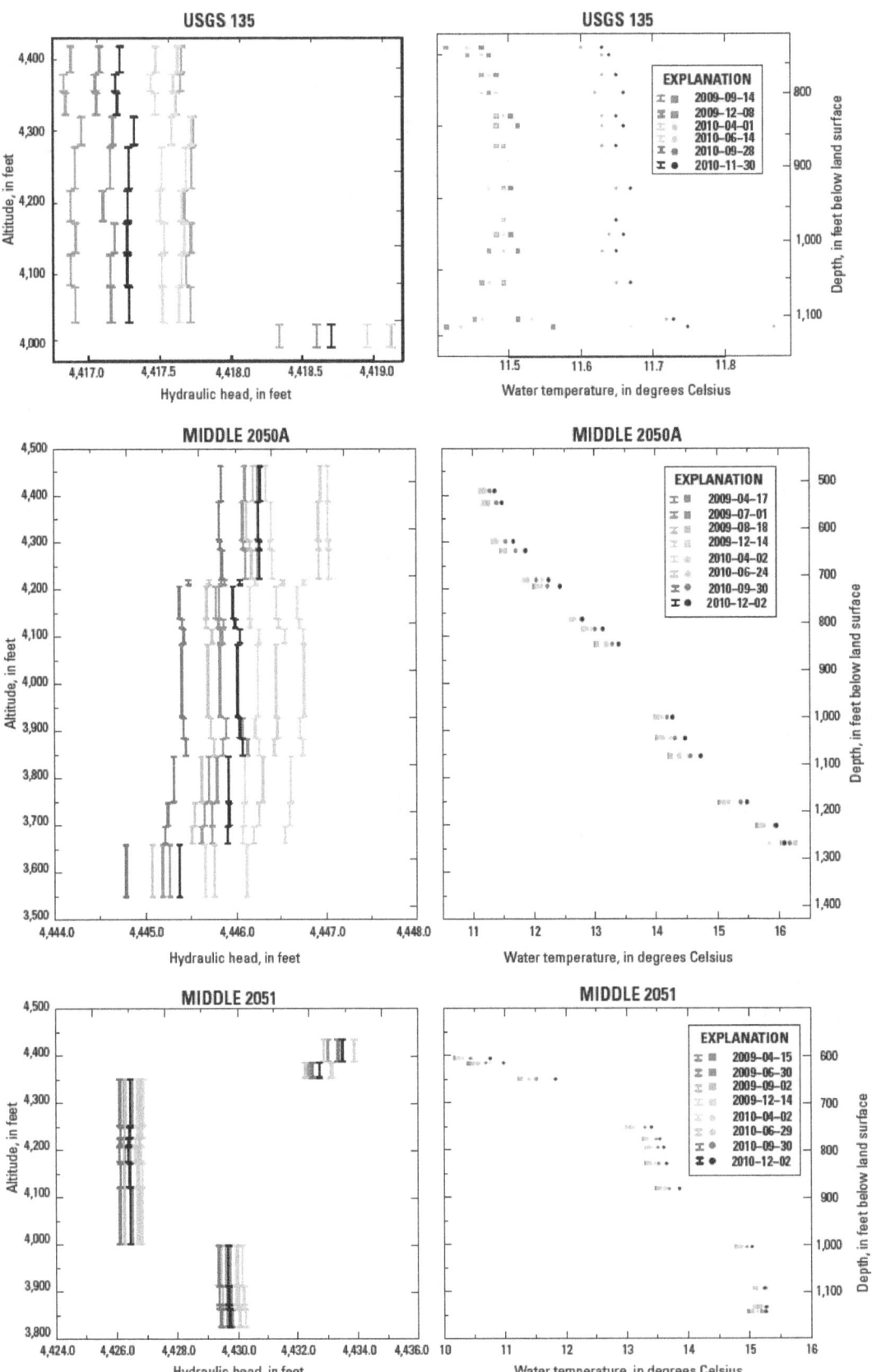

Figure 7.—Continued.

Table 3. Minimum Pearson correlation coefficients for hydraulic head and temperature profiles at selected boreholes, Idaho National Laboratory, Idaho, 2009–10.

[**Local name:** Local well identier used in this study]

Local name	Pearson correlation coefficients	
	Hydraulic head	Temperature
USGS 103	0.52	0.97
USGS 105	0.64	0.96
USGS 108	1.00	0.99
USGS 132	0.84	0.99
USGS 133	1.00	0.97
USGS 134	0.82	1.00
USGS 135	1.00	-0.57
MIDDLE 2050A	0.93	0.99
MIDDLE 2051	1.00	1.00

Minimum PCC values for head profiles ranged from 0.52 at borehole USGS 103 to 1.00 at boreholes USGS 108, USGS 133, USGS 135, and MIDDLE 2051 (table 3). In three boreholes (USGS 103, 105, and 132), the minimum PCC was less than 0.9, and is attributed to small vertical head differences and a relative uncertainty for head differences between adjacent zones of ±0.1 ft; under these circumstances, measurement error can produce a lower PCC where a strong correlation exists. Additionally, borehole USGS 134 has a minimum PCC of 0.82, partially because of the large head difference between paired ports in monitoring zone 15 (fig. 6). This head difference is believed to suggest evidence of a pressure response to mountain front recharge events. All other MLMS boreholes resulted in minimum PCC values greater than or equal to 0.90, which suggests a strong positive correlation among head profiles.

Calculation of the minimum PCC values for temperature profiles ranged from -0.57 in borehole USGS 135 to 1.00 in boreholes USGS 134 and MIDDLE 2051 (table 3). The minimum PCC values for temperature profiles were relatively strong, except for borehole USGS 135. Temperature in borehole USGS 135 indicates groundwater is slightly warmer by about 0.2°C during third and fourth quarter measurements for 2010; however, the source of the warmer water is not well understood. The negative PCC suggests groundwater temperature at this location has not reached equilibrium.

For eight of the nine boreholes, head and temperature were examined for the June 2010 profile. The exception was borehole USGS 108, where the September 2010 profile was examined because the installation was not completed in June. The profiles for all nine MLMSs are presented with their corresponding borehole information in figures 8–16. Borehole information includes four geophysical traces, a lithology log, and a multilevel completion log. The information for six of the wells (USGS 103, USGS 132, USGS 133, USGS 134,

MIDDLE 2050A, and MIDDLE 2051) also was previously given in Fisher and Twining (2011), but are presented here for comparison with new head and temperature profiles from June 2010. The geophysical traces are:

1. Natural gamma is a measure of the gamma radiation emitted by the naturally occurring radioisotopes within the rock material composing the borehole wall. Elevated natural gamma readings typically indicate the presence of a sedimentary layer.

2. Neutron is a measure of the hydrogen content of the rock, which, when saturated, is directly related to the porosity of the porous medium. A high neutron porosity indicates the presence of highly fractured basalt or sediment; whereas, a low neutron porosity would indicate an area of dense basalt.

3. Left and right caliper uses three extendable spring loaded arms to measure drill-hole diameter with an accuracy of ±0.15 in. Changes in the regular drill-hole diameter may be due to collapse of the loose or highly fractured rock formations—areas unsuitable for packer placement.

4. Short-spaced and long-spaced gamma-gamma dual density, also known as the induced gamma-density, is a measure of the bulk density of a rock material near a borehole wall. The bulk density of a rock material is inversely related to its porosity.

All geophysical traces, along with the borehole video and a visual inspection of the core, were used to construct the generalized lithology log for each MLMS borehole (figs. 8–16; appendix E). Generalized lithology for six of the wells (USGS 103, USGS 132, USGS 133, USGS 134, MIDDLE 2050A, and MIDDLE 2051) previously was given in Fisher and Twining (2011) and is not included in appendix E. The lithology logs for all boreholes were described using three lithologic units: (1) dense basalt, a rock material of moderate to low horizontal hydraulic conductivity and low to very low vertical hydraulic conductivity; (2) fractured basalt, a rock material of high to very high hydraulic conductivity; and (3) sediment, a fine-grained sand and silt mixture of very low hydraulic conductivity. The percentage of lithologic composition in each borehole is provided in table 4. The reported effective hydraulic conductivity for the basalt and interbedded sediment that compose the ESRP aquifer at or near the INL ranges from about 1.0×10^{-2} to 3.2×10^{4} ft/d (Anderson and others, 1999). Reported porosity of the aquifer based on a cumulative distribution curve for more than 1,500 individual cores showed that the central 80 percent of samples had porosities between 0.08 and 0.25 (Knutson and others, 1992, figs. 4–10).

The multilevel completion(s) are displayed for each borehole (figs. 8–16) and include the location of measurement ports, packers, and monitoring zones. Measurement ports and monitoring zones are labeled using 'P' and 'Z', respectively,

Table 4. Lithologic composition in selected boreholes, Idaho National Laboratory, Idaho, 2009–10.

[**Local name** is the local well identifier used in this study]

	Boreholes		
	Percentage of lithologic unit		
Local name	Dense basalt	Fractured basalt	Sediment
USGS 103	55	41	4
USGS 105	45	52	3
USGS 108	49	46	5
USGS 132	35	63	2
USGS 133	56	34	10
USGS 134	46	51	3
USGS 135	52	45	3
MIDDLE 2050A	51	35	14
MIDDLE 2051	64	33	3

followed by a unique index number that increases with decreasing depth. For example, P3 is the third measurement port from the bottom of the hole, and Z4 is the fourth monitoring zone from the bottom.

The shapes of the head profiles were analyzed using major head inflections for the June 2010 dataset in eight of the nine multilevel monitoring wells. The exception was USGS 108, for which the September 2010 dataset was used. These inflections were identified using the difference between head measurements of adjacent monitoring zones. Head inflections were considered major where head differences exceeded ±0.1 ft, the relative uncertainty for head differences between adjacent zones. The head inflections were labeled using 'H' followed by a unique index number that increases with decreasing depth. For example, H1 identifies the vertical location of the first head inflection from the bottom of the hole. Vertical head gradients were calculated across the 3.0-ft-thick inflated packer length that separates monitoring zones (table 5; appendix F).

Table 5. Vertical hydraulic gradients at major inflection points for depth interval and hydraulic head profiles, Idaho National Laboratory, Idaho, June 2010.

[Major inflection points were identified using the differences between hydraulic head (head) measurements of adjacent monitoring zones. Head inflections were considered major where head differences exceeded the relative uncertaining for head differences between adjacent zones, ±0 1 foot. **Local name** is the local well identifier used in this study. Location of boreholes is shown in figure 1. **Inflection index No :** Identifier used to locate major inflection points in the head profile. **Zone No.:** Identifiers used to locate monitoring zones. **Port No :** Idenifiers used to locate port couplings. **Depth interval:** Depth to the bottom and top of the inflated packer separating the adjacent monitoring zone. **Hydraulic head:** Negative (-) and positive values indicate heads decreasing and increasing with depth, repectively. **Abbreviations:** ft bls, feet below land surface; ft, foot; ft/ft, foot per foot]

				Boreholes				
				Depth interval			Hydraulic head	
Local name	Inflection index No.	Zone No.	Port No.	Bottom (ft bls)	Top (ft bls)	Length (ft)	Difference (ft)	Gradient (ft/ft)
USGS 103	H1	15, 16	19–21	766.9	763.9	3.0	-0.4	-0.1
	H2	16, 17	21–23	694.3	691.3	3.0	0.3	0.1
USGS 105	H1	10, 11	13–15	865.3	862.3	3.0	0.2	0.1
USGS 108	H1	2, 3	3, 4	1,121.6	1,118.6	3.0	-0.2	-0.1
	H2	6, 7	8–10	906.7	903.7	3.0	0.3	0.1
	H3	9, 10	12–14	791.4	788.4	3.0	0.3	0.1
USGS 133	H1	4, 5	5, 6	685.5	682.5	3.0	-4.8	-1.6
	H2	5, 6	6, 7	618.2	615.2	3.0	-0.6	-0.2
	H3	6, 7	7–9	593.7	590.7	3.0	-0.8	-0.3
USGS 134	H1	9, 10	11–13	690.9	687.9	3.0	-0.7	-0.2
	H2	10, 11	13–14	664.9	661.9	3.0	0.8	0.3
	H3	14, 15	18–20	592.8	589.8	3.0	-0.2	-0.1
USGS 135	H1	1, 2	1–3	1,105.6	1,102.6	3.0	1.5	0.5
MIDDLE 2050A	H1	1, 2	1, 2	1,267.5	1,264.5	3.0	-0.4	-0.1
	H2	11, 12	11, 12	706.4	703.4	3.0	-0.4	-0.1
MIDDLE 2051	H1	4, 5	4, 5	1,002.2	999.2	3.0	3.3	1.1
	H2	10, 11	10, 11	646.7	643.7	3.0	-6.3	-2.1
	H3	11, 12	11, 12	612.2	609.2	3.0	-0.8	-0.3

USGS 103

Borehole USGS 103 was established along the southern boundary of the INL about 5.5 mi south of the Central Facilities Area (CFA) (fig. 1). The land-surface altitude and the estimated altitude of the base of the aquifer at this location are 5,007.42 and 2,470 ft, respectively (table 1). The MP55 system extends to a depth of 1,279.4 ft bls and includes 23 measurement ports and 17 monitoring zones; 6 of these zones contain paired ports. Zone lengths range from 6.6 to 69.7 ft (fig. 8; table 2).

Two inflections were identified in the borehole USGS 103 head profile (fig. 8): (1) H1, located across the 3-ft packer separating zones 15 and 16 with a -0.4 ft (downward) head loss; and (2) H2, located between zones 16 and 17 with a 0.3-ft head gain (fig. 8, table 5). The H1 and H2 inflections occur where the MLMS enters steel casing near 760 ft bls; therefore, change in head pressure is likely the result of borehole construction and not attributed to aquifer response. This anomalous head inflection near zone 16 only occurs in two of eight profile events (fig. 7). The borehole, starting at the bottom near zone 1 up to and including zone 15, shows a high degree of vertical connectivity among adjacent fracture sets. The range of head in the profile was relatively small at 0.6 ft and indicates flow that is dominantly horizontal. Sediment layers in the borehole had no apparent effect on head. Water temperatures in the borehole USGS 103 temperature profile ranged from 12.4 to 13.2°C and averaged 12.6°C (fig. 8). Temperature generally decreases with depth in the upper part of the profile and increases with depth in the lower part with a transition near 1,100 ft bls.

USGS 105

Borehole USGS 105 was established along the southern boundary of the INL about 5.9 mi south of the CFA (fig. 1). The land-surface altitude and the estimated altitude of the base of the aquifer at this location are 5,095.12 and 2,540 ft, respectively (table 1). The borehole was air-rotary drilled in 1980 to a depth of 801 ft bls and completed as an 8-in. open-interval monitoring well. In 2008, the borehole was rotary drilled to 1,409 ft bls and then reamed to 1,300 ft bls to allow installation of a MP55 MLMS. Prior to reaming, the 5-in. casing was permanently set to a depth of 801 ft bls to allow proper packer inflation during the May 2008 MLMS installation; the MP55 system required a borehole diameter not to exceed 6.25 in. Perforations in the 6-in. casing were placed within monitoring zone 13 (fig. 9) to allow measurement ports 17 and 18 access to free moving groundwater, although the vertical isolation of groundwater within zones 12 and 13 was compromised by the annular space between the 5-in. casing and the 8-in. borehole wall. Therefore, measurements collected from ports 16, 17, and 18 reflect a vertically averaged value of head for the interval between the water table and the bottom of zone 12. The MP55 system extends to a depth of 1,290.1 ft bls and includes 18 measurement ports and 13 monitoring zones; 5 of these zones contain paired ports. Zone lengths range from 10.9 to 72.5 ft (fig. 9; table 2).

The lithology log of borehole USGS 105 (fig. 9; appendix E) shows units that range from 3 to 25 ft for dense basalt, 1 to 93 ft for fractured basalt, and 2 to 7 ft for sediment. The composition of lithologic units in the borehole is 45 percent dense basalt, 52 percent fractured basalt, and 3 percent sediment (table 4). Six sediment layers are in the borehole at depths of 816, 873, 879, 993, 1,132, and1,252 ft bls (fig. 9). Sediment recovered during drilling was described as eolian deposits of fine- grained sand with silt. The layers of fractured and dense basalt are numerous and seem to be well distributed throughout the borehole.

A single inflection was identified in the borehole USGS 105 head profile (fig. 9): H1, located across the 3-ft packer separating zones 10 and 11 with a 0.2 ft head gain (fig. 9, table 5). The H1 inflection coincides with layers of low-permeability sediment —where the sediment at this location obstructs the vertical connectivity between adjacent fracture sets. The range of head in the USGS 105 profile was relatively small at 0.4 ft (table 6) and indicates flow that is dominantly horizontal. Sediment layers in the borehole show minimal effect on hydraulic head. Water temperatures in the borehole USGS 105 temperature profile ranged from 12.4 to 13.0°C and averaged 12.7°C (fig. 9). A reversal in the temperature gradient occurred at about 1,000 ft bls, with the coldest water temperature reported in the lowest ports (12.4 to 12.5°C at ports 1 through 4). Generally, water temperature increases with depth; however, this trend mirrors that seen in borehole USGS 132 (fig. 7).

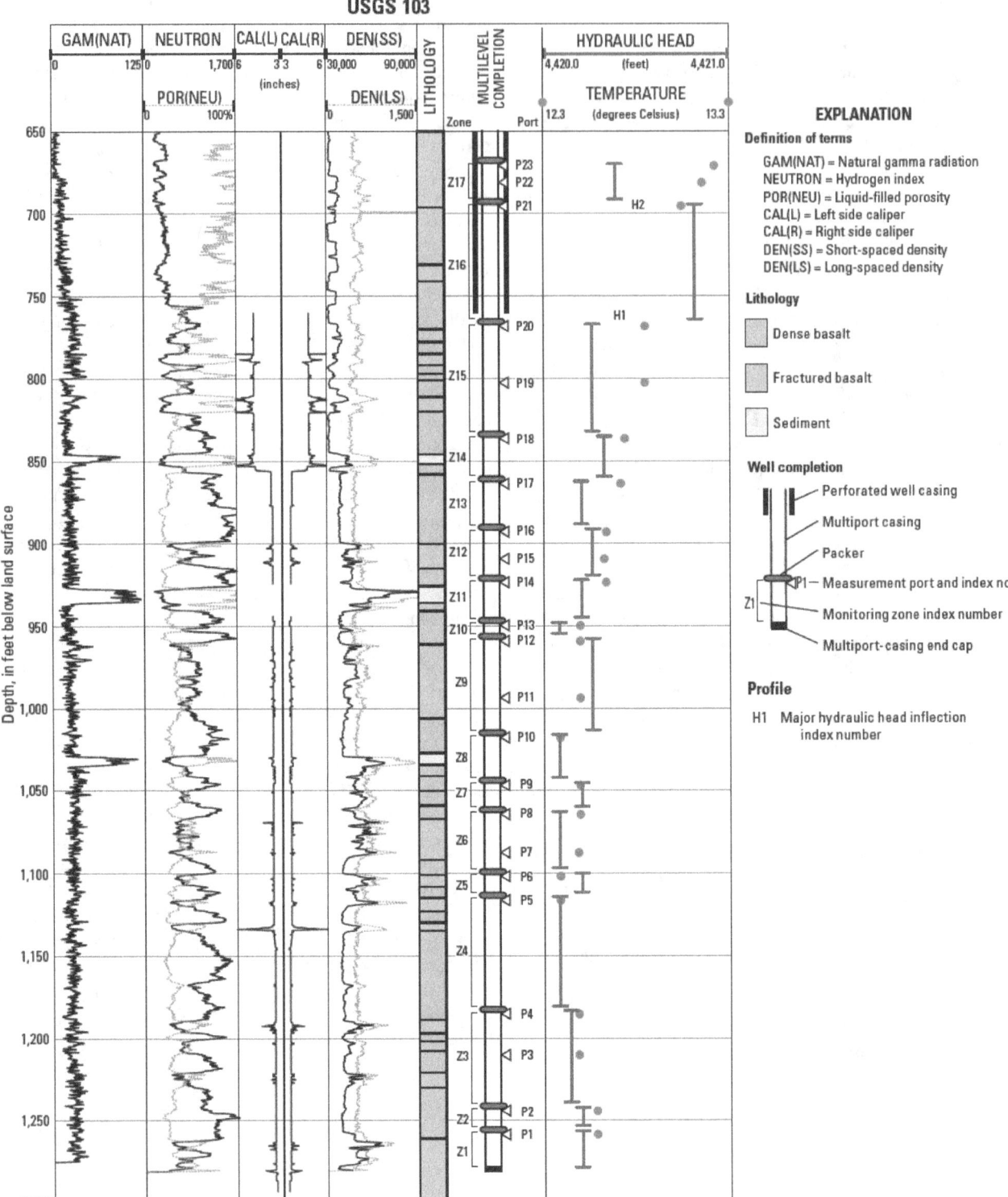

Figure 8.   Geophysical traces of natural gamma, neutron, caliper, and gamma-gamma dual density; lithology log; multilevel completion; and hydraulic head and water temperature profiles for borehole USGS 103, Idaho National Laboratory, Idaho, June 2010.

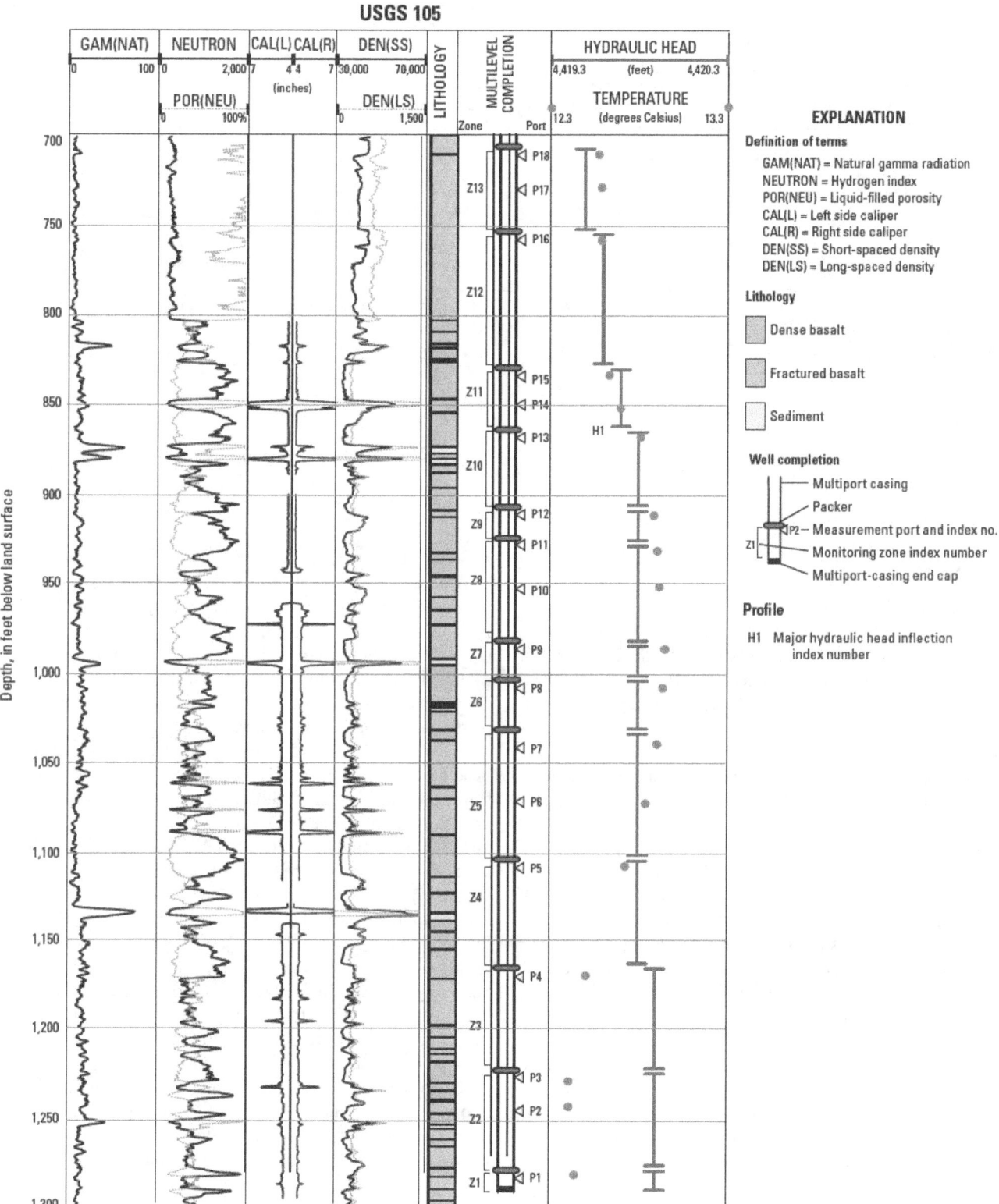

Figure 9. Geophysical traces of natural gamma, neutron, caliper, and gamma-gamma dual density; lithology log; multilevel completion; and hydraulic head and water temperature profiles for borehole USGS 105, Idaho National Laboratory, Idaho, June 2010.

Table 6.   Summary of depth range, hydraulic head statistics, fluid temperature statistics, water-level depth, and saturated thickness of the aquifer at each borehole, Idaho National Laboratory, Idaho, June 2010.

[**Local name** is the local well identifier used in this study. Location of boreholes is shown in figure 1. **Depth interval** is measured from the top of the uppermost monitoring zone to the bottom of the lowest zone in feet (ft). **Hydraulic head statistics** include the mean in feet above National Geodetic Vertical Datum of 1929 (NGVD29) and the range in ft. **Fluid temperature statistics** include the mean and the range in degrees Celsius (C). **Water-level depth** is in feet below land surface (ft bls). **Saturated aquifer thickness** in feet is determined from subtracting the water-level depth from the aquifer thickness (appendix A)]

		Boreholes					
Local name	Interval length (ft)	Hydraulic head statistics		Fluid temperature statistics		Water-level depth (ft bls)	Saturated aquifer thickness (ft)
		Mean (ft amsl)	Range (ft)	Mean (°C)	Range (°C)		
USGS 103	610	4,420.2	0.6	12.6	0.8	584.63	1,950
USGS 105	583	4,419.8	0.4	12.7	0.5	675.32	1,950
USGS 108	550	4,419.2	0.9	12.6	0.3	612.16	1,950
USGS 132	590	4,420.0	0.3	11.9	1.9	607.10	1,880
USGS 133	318	4,460.6	6.2	11.6	1.2	427.61	500
USGS 134	333	4,453.6	0.9	13.8	2.1	513.70	500
USGS 135	410	4,417.7	1.6	11.5	0.2	718.24	1,870
MIDDLE 2050A	913	4,446.5	1.1	13.2	5.2	479.94	660
MIDDLE 2051	615	4,428.9	7.2	13.3	4.5	569.48	1,160

USGS 108

Borehole USGS 108 was established along the southern boundary of the INL about 6.3 mi south of the CFA (fig. 1). The land-surface altitude and the estimated altitude of the base of the aquifer at this location are 5,031.36 and 2,495 ft, respectively (table 1). The borehole was air-rotary drilled in 1980 to a depth of 760 ft bls and completed as an 8-in. open interval monitoring well. In 2008, the borehole was deepened to 1,218 ft bls using rotary coring and then reamed to allow installation of a MP55 MLMS. Prior to reaming, the 5-in. casing was permanently set to a depth of 760 ft bls to allow proper packer inflation during the August 2010 MLMS installation; the MP55 system required a borehole diameter not to exceed 6.25 in. Perforations in the 5-in. casing were placed within monitoring zone 11 (fig. 10) to allow measurement ports 15 and 16 access to free moving groundwater, although the vertical isolation of groundwater within zones 10 and 11 was compromised by the annular space between the 5-in. casing and the 8-in. borehole wall. Therefore, measurements collected from ports 14, 15, and 16 reflect a vertically averaged value of head for the interval between the water table and the bottom of zone 10. The MP55 system extends to a depth of 1,191.9 ft bls and includes 16 measurement ports and 11 monitoring zones; 5 of these zones contain paired ports. Zone lengths range from 31.0 to 106.6 ft (fig. 10; table 2).

The lithology log of borehole USGS 108 (fig. 10; appendix E) shows units that range in length from 3 to 33 ft for dense basalt, 1 to 25 ft for fractured basalt, and 3 to 7 ft for sediment. The composition of lithologic units in the borehole is 49 percent dense basalt, 46 percent fractured basalt, and 5 percent sediment (table 4). Six sediment layers were identified near depths of 854, 878, 888, 1,000, 1,103, and 1,169 ft bls (fig. 10). Sediment recovered during drilling was described as eolian deposits of fine-grained sand and silt. Layers of fractured and dense basalt are numerous and seem to be well distributed throughout the borehole.

Three inflections were identified in the USGS 108 head profile (fig. 10): (1) H1, located across the 3-ft packer separating zones 2 and 3 with a -0.2 ft downward head loss; (2) H2, located between zones 6 and 7 with a 0.3-ft head gain; and (3) H3, located between zones 9 and 10 with a 0.3-ft head gain (fig. 10, table 5). The H1 and H2 inflections coincide with layers of low-permeability sediment—where the sediment at this location is believed to obstruct the vertical connectivity between adjacent fracture sets. The H3 inflection occurs within the upper part of the MLMS, where zone 10 is a long vertically averaged interval (fig. 10). Borehole construction can not be ruled out as the controlling factor for the H3 inflection, however; dense basalt strata within the bottom of zone 10 may result in a pressure change. The range of head in the profile was 0.9 ft (table 6) and indicates flow that is dominantly horizontal. Sediment layers in the borehole had minimal effect on head change. Water temperatures in the borehole USGS 108 temperature profile ranged from 12.4 to 12.7°C and averaged 12.6°C (fig. 10). Temperature generally increased with depth within a very small range at 0.3°C.

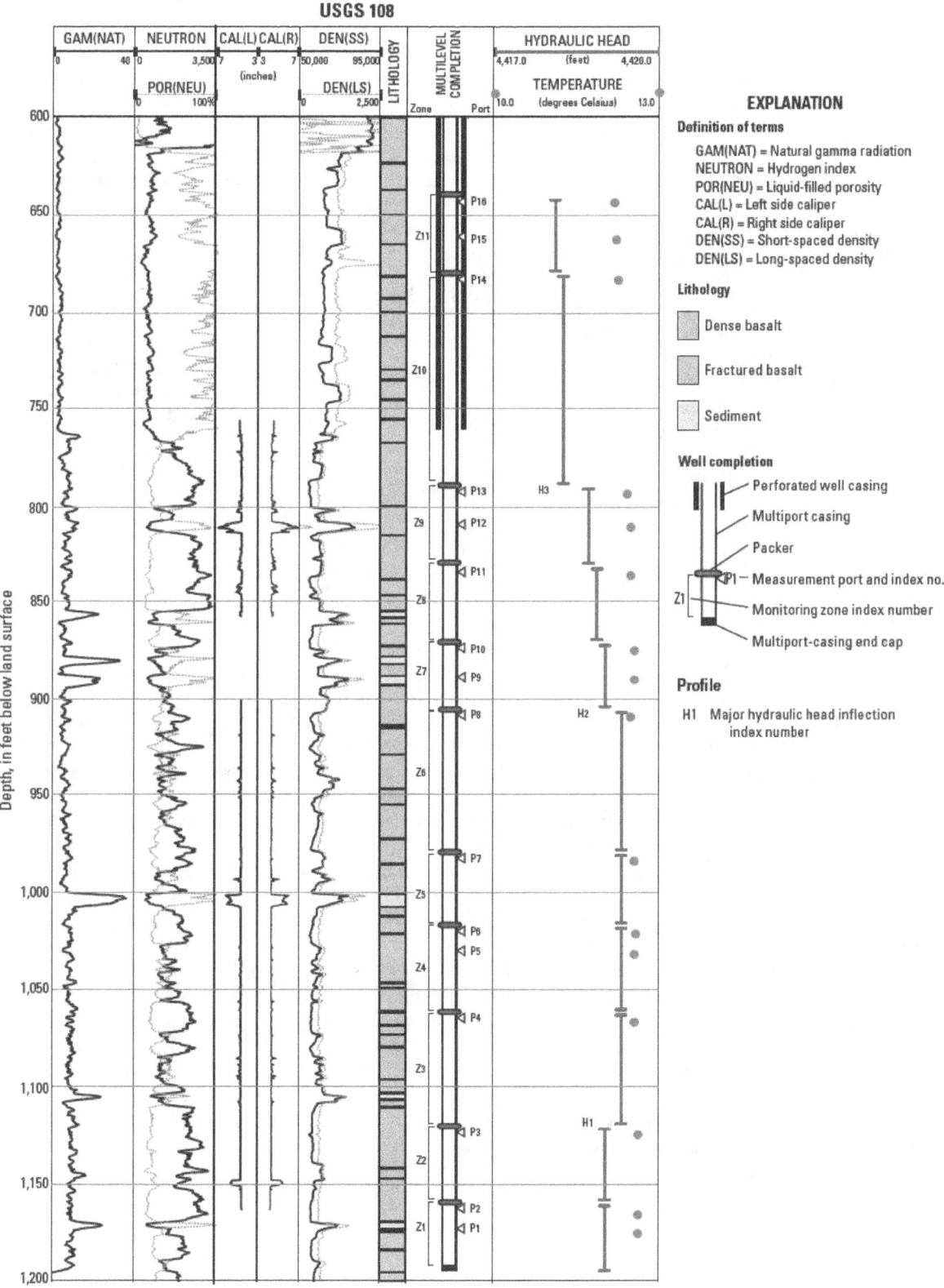

Figure 10. Geophysical traces of natural gamma, neutron, caliper, and gamma-gamma dual density; lithology log; multilevel completion; and hydraulic head and water temperature profiles for borehole USGS 108, Idaho National Laboratory, Idaho, September 2010.

USGS 132

Borehole USGS 132 was established about 0.9 mi south of the Radioactive Waste Management Complex (RWMC) (fig. 1). The land-surface altitude and the estimated altitude of the base of the aquifer at this location are 5,028.60 and 2,540 ft, respectively (table 1). The MP55 system extends to a depth of 1,213.6 ft bls and has a total of 23 measurement ports and 17 monitoring zones; 6 of these zones contain paired ports. Zone lengths range from 5.2 to 85.3 ft (fig. 11; table 2).

No inflections were identified in the USGS 132 head profile (fig. 11). The absence of inflections and the abundance of fractured basalt indicate a high degree of vertical connectivity among adjacent fracture sets. The range of head in the profile was relatively small at 0.3 ft (table 6) and indicates flow that is dominantly horizontal. Water temperatures in the USGS 132 temperature profile (fig. 11) ranged from 10.6 to 12.5°C and averaged 11.9°C. A reversal in the temperature gradient occurs at an inflection point about 850 ft bls, with temperatures decreasing with depth below the inflection point and increasing with depth above.

USGS 133

Borehole USGS 133 was established about 1.9 mi north of the Idaho Nuclear Technology and Engineering Center (INTEC) (fig. 1). The land-surface altitude and the estimated altitude of the base of the aquifer at this location are 4,890.12 and 3,960 ft, respectively (table 1). The MP55 system extends to a depth of 766.4 ft bls and includes 13 measurement ports and 10 monitoring zones; 3 of the zones contain paired ports. Zone lengths ranged from 6.8 to 64.3 ft (fig. 12; table 2).

Three major inflections were identified in the borehole USGS 133 head profile: (1) H1, located across the 3-ft packer separating zones 4 and 5 with a -4.8 ft downward head loss; (2) H2, located between zones 5 and 6 with a -0.6 ft downward head loss; and (3) H3, located between zones 6 and 7 with a -0.8 ft downward head loss (fig. 12; table 5). The source of inflections H1, H2, and H3 is the 33- ft sediment layer—where the sediment behaves as a confining layer, isolating fractured flow within the upper and lower parts of the aquifer. The integrity of this confining layer potentially was compromised by allowing downward flow to bypass the sediment layer through the open borehole of monitoring zone 5, as first described by Fisher and Twining (2011). The H1 inflection directly coincides with this sediment layer and has a downward vertical hydraulic gradient of -1.6 ft/ft. Vertical gradients above the large sediment layer are somewhat less in magnitude at -0.2 and -0.3 ft/ft for the H2 and H3 inflections (negative values indicate a downward gradient), respectively, and reflect a transition from low- permeability sediment and dense basalt to high-permeability fractured basalt. Head differences below the H1 inflection and above the H3 inflection remained relatively small, indicating flow that is dominantly horizontal. Water temperatures in the borehole USGS 133 temperature profile (fig. 12) gradually increased with depth, ranged from 11.0 to 12.2°C, and averaged 11.6°C.

USGS 134

Borehole USGS 134 was established about 1.8 mi northwest of the Advanced Test Reactor (ATR) (fig. 1). The land-surface altitude and the estimated altitude of the base of the aquifer at this location are 4,968.84 and 3,960 ft, respectively (table 1). The MP38 system extends to a depth of 886.8 ft bls and includes 20 measurement ports and 15 monitoring zones; 5 of these zones contain paired ports. Zone lengths range from 5.8 to 36.0 ft (fig. 13; table 2).

Three major inflections were identified in the borehole USGS 134 head profile: (1) H1, located across the 3-ft packer separating zones 9 and 10 with a -0.7 ft downward head loss; (2) H2, located between zones 10 and 11 with a 0.8 ft upward head gain; and (3) H3, located between zones 14 and 15 with a -0.2 ft downward head loss (fig. 13; table 5). The vertical hydraulic gradients for the H1, H2, and H3 inflections are -0.2, 0.3, and -0.1 ft/ft (positive values indicate an upward gradient), respectively. The response of inflections H1 and H2 result from a localized head increase in zone 10. This head increase probably is hydraulically isolated in zone 10 because of the relatively small 0.1-ft head difference between zones 9 and 11, and possibly because of fractures in this zone that are poorly connected to the larger hydraulic system. The H3 inflection occurs near the top of the aquifer; fractures within zone 15 appear to be partially isolated by dense basalt. The location of the H3 inflection and proximity to mountain fronts (fig. 1) may suggest the upper aquifer is responding to periodic tributary valley subsurface inflow. Excluding zone 10, the variability in head was small, with a range of 0.9 ft. Vertical head remained relatively stable below H1 and gradually decreased with depth above H2. Water temperatures in the borehole USGS 134 temperature profile (fig. 13) gradually increased with depth, ranged from 12.8 to 14.9°C, and averaged 13.8°C.

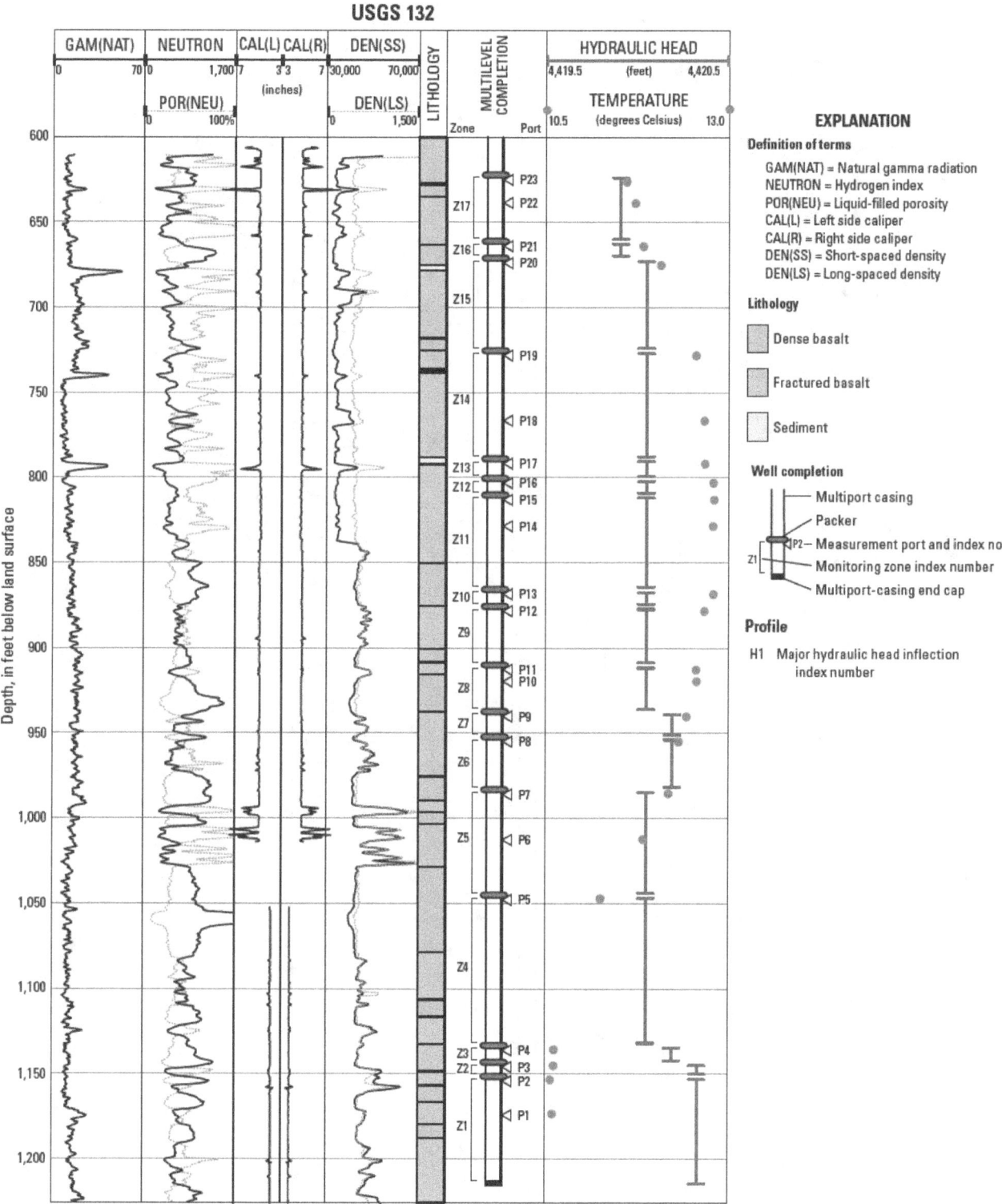

Figure 11. Geophysical traces of natural gamma, neutron, caliper, and gamma-gamma dual density; lithology log; multilevel completion; and hydraulic head and water temperature profiles for borehole USGS 132, Idaho National Laboratory, Idaho, June 2010.

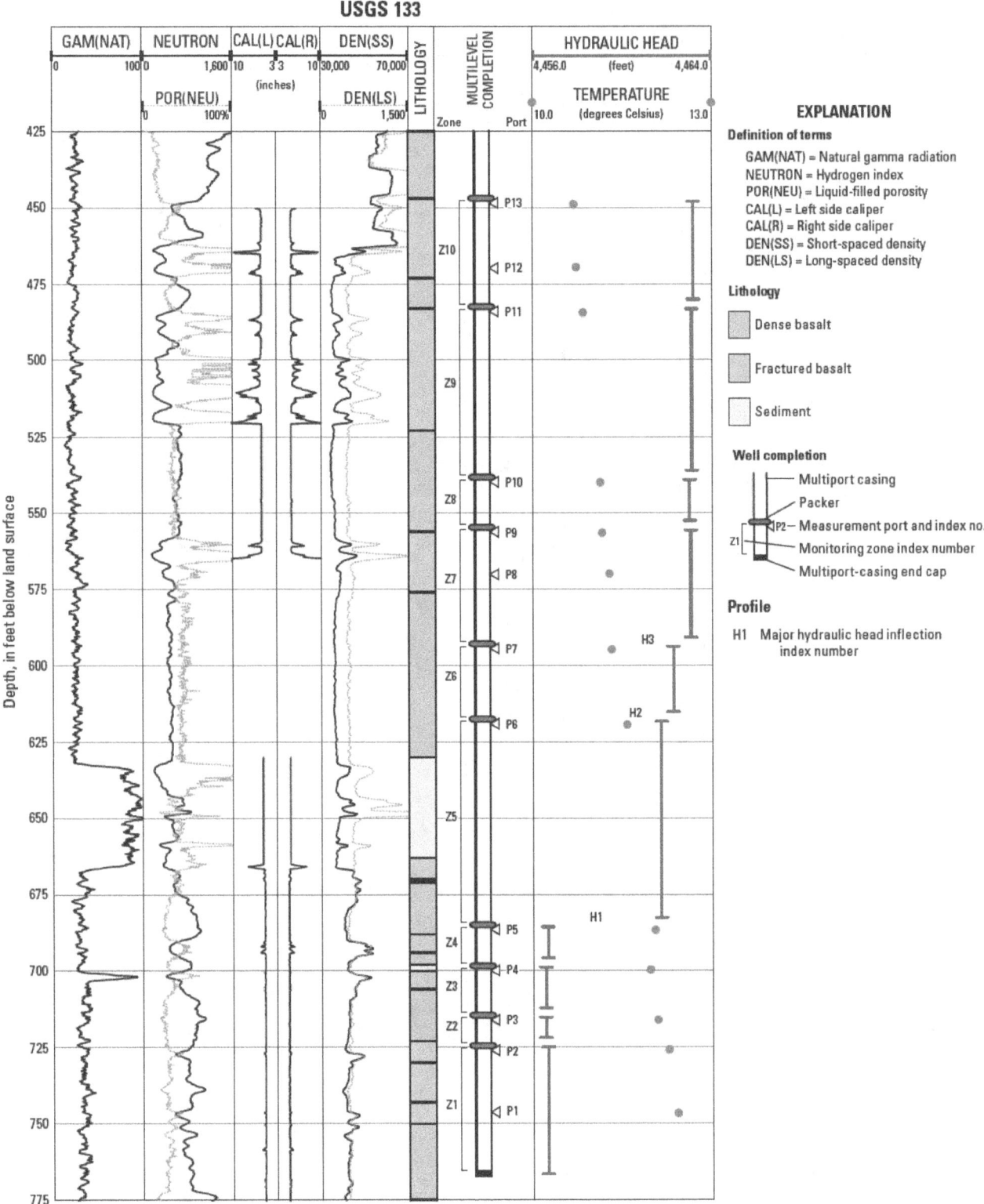

Figure 12.   Geophysical traces of natural gamma, neutron, caliper, and gamma-gamma dual density; lithology log; multilevel completion; and hydraulic head and water temperature profiles for borehole USGS 133, Idaho National Laboratory, Idaho, June 2010.

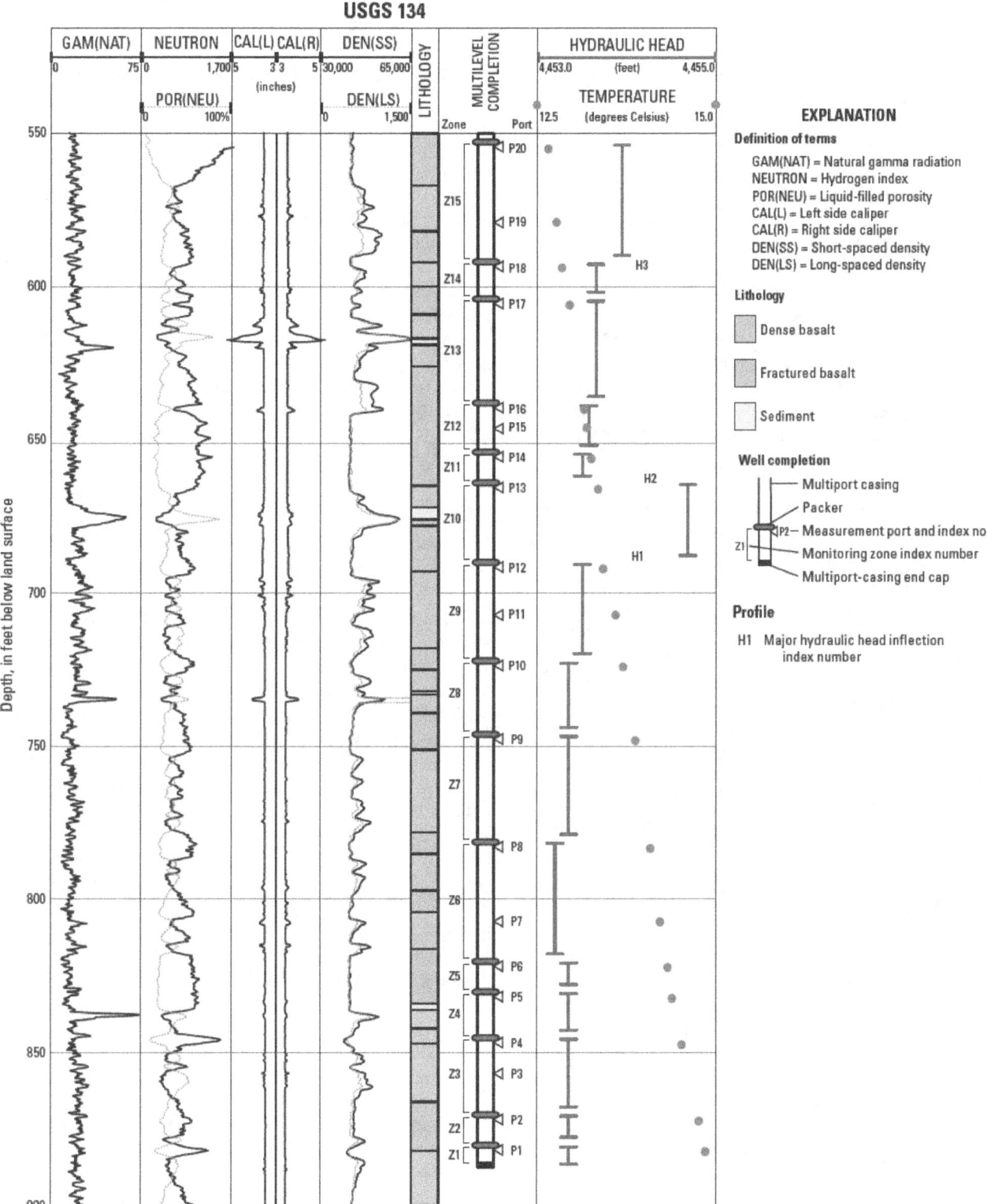

Figure 13. Geophysical traces of natural gamma, neutron, caliper, and gamma-gamma dual density; lithology log; multilevel completion; and hydraulic head and water temperature profiles for borehole USGS 134, Idaho National Laboratory, Idaho, June 2010.

USGS 135

Borehole USGS 135 was completed along the southwestern corner on the INL, about 5.0 mi southwest of the RWMC (fig. 1). The land-surface altitude and the estimated altitude of the base of the aquifer at this location are 5,135.94 and 2,675 ft, respectively (table 1). In 2007, the borehole was continuously cored to a depth of 1,198 ft bls, and in July 2009, an MP55 MLMS was installed to a depth of 1,136.6 ft bls. The borehole was completed with 14 measurement ports and 10 monitoring zones; 4 of these zones contain paired ports. Zone lengths range from 21.6 to 56.1 ft (fig. 14; table 2).

The lithology log of borehole USGS 135 (fig. 14; appendix E) shows units that range in length from 2 to 49 ft for dense basalt, 1 to 38 ft for fractured basalt, and 1 to 5 ft for sediment. The composition of lithologic units in the borehole is 52 percent dense basalt, 45 percent fractured basalt, and 3 percent sediment (table 4). Six sediment layers in the borehole are at approximate depths of 854, 856, 871, 919, 960, and 972 ft bls (fig. 14). Sediment recovered during drilling was described as eolian deposits of silt with clay. Layers of fractured and dense basalt are numerous, with high concentrations of fractured basalt starting near 875 ft bls.

A single inflection was identified in the borehole USGS 135 head profile: (1) H1, located across the 3-ft packer separating zones 1 and 2 with a 1.5 ft head gain (fig. 14; table 5). The H1 inflection has a vertical hydraulic gradient of 0.5 ft/ft and coincides with layers of fractured basalt (starting near 1,115 ft bls) overlain by a 50-ft-thick layer of dense basalt. The low permeability associated with dense basalt creates a hydraulic disconnect between the high-permeability fractured basalt, resulting in a measurable head gain. Excluding zone 1, the variability in head was small and vertical head remained stable above H1. Water temperatures in the USGS 135 temperature profile (fig. 14) gradually increased with depth, ranged from 11.4 to 11.7°C, and averaged 11.5°C.

MIDDLE 2050A

Borehole MIDDLE 2050A was established about 0.8 mi west of the INTEC and 0.22 mi southeast of the Big Lost River (fig. 1). The land-surface altitude and the estimated altitude of the base of the aquifer at this location are 4,928.22 and 3,790 ft, respectively (table 1). The MP55 system extends to a depth of 1,377.6 ft bls and includes 15 measurement ports and 15 monitoring zones. Zone lengths range from 10.2 to 152.6 ft (fig. 15; table 2).

Two major inflections were identified in the MIDDLE 2050A head profile: (1) H1, located across the 3-ft packer separating zones 1 and 2 with a -0.4 ft (downward) head loss; and (2) H2, located between zones 11 and 12 with a -0.4 ft

head loss (fig. 15, table 5). The vertical hydraulic gradient of the H1 inflection was -0.1 ft/ft and coincides with the 100-ft-thick layer of sediment. The head in zone 1 likely is controlled by the units of fractured basalt above and below this sediment layer, where flow is allowed to bypass the low-permeability sediment through the open borehole (Fisher and Twining, 2011). The inflection near H2 resulted in a vertical hydraulic gradient of -0.1 ft/ft, and coincides with layers of low-permeability sediment—where the sediment at these locations obstructs the vertical connectivity between adjacent fracture sets. Head differences between the major inflections remained relatively small and indicated flow that is dominantly horizontal; however, head decreased incrementally with depth in the borehole, creating the potential for downward flow. Water temperatures in the borehole MIDDLE 2050A temperature profile gradually increased with depth, ranged from 11.1 to 16.3°C, and averaged 13.2°C (fig. 15).

MIDDLE 2051

Borehole MIDDLE 2051 was established about 2.75 mi northeast of the RWMC and 0.16 mi southeast of the Big Lost River (fig. 1). The land-surface altitude and the estimated altitude of the base of the aquifer at this location are 4,997.31 and 3,270 ft, respectively (table 1). The MP55 system extends to a depth of 1,176.5 ft bls and includes 12 measurement ports and 12 monitoring zones. Zone lengths range from 6.8 to 119.8 ft (fig. 16; table 2).

Three major inflections were identified in the borehole MIDDLE 2051 head profile: (1) H1, located across the 3-ft packer separating zones 4 and 5 with a 3.3 ft downward head gain; (2) H2, located between zones 10 and 11 with a -6.3 ft head loss; and (3) H3, located between zones 11 and 12 with a -0.8 ft head loss (fig. 16; table 5). The H1 inflection has a vertical hydraulic gradient of 1.1 ft/ft and coincides with layers of fractured basalt separated by a 5-ft-thick layer of sediment and a 75-ft-thick layer of dense basalt (zone 5). The low permeability associated with these layers creates a hydraulic disconnect between the high-permeability fracture sets above and below zone 5. The H2 and H3 inflections have vertical hydraulic gradients of -2.1 and -0.3 ft/ft, respectively, and also result from a restriction in vertical connectivity—where fracture sets coinciding with zones 10 and 12 are hydraulically isolated from one another by two layers of low-permeability sediment (5- and 7-ft thick) within zone 11. Head differences below the H1 inflection and between the H1 and H2 inflections remained relatively small, indicating flow that is dominantly horizontal. Water temperatures generally decreased with depth, ranged from 10.5 to 15.0 °C, and averaged 13.3°C.

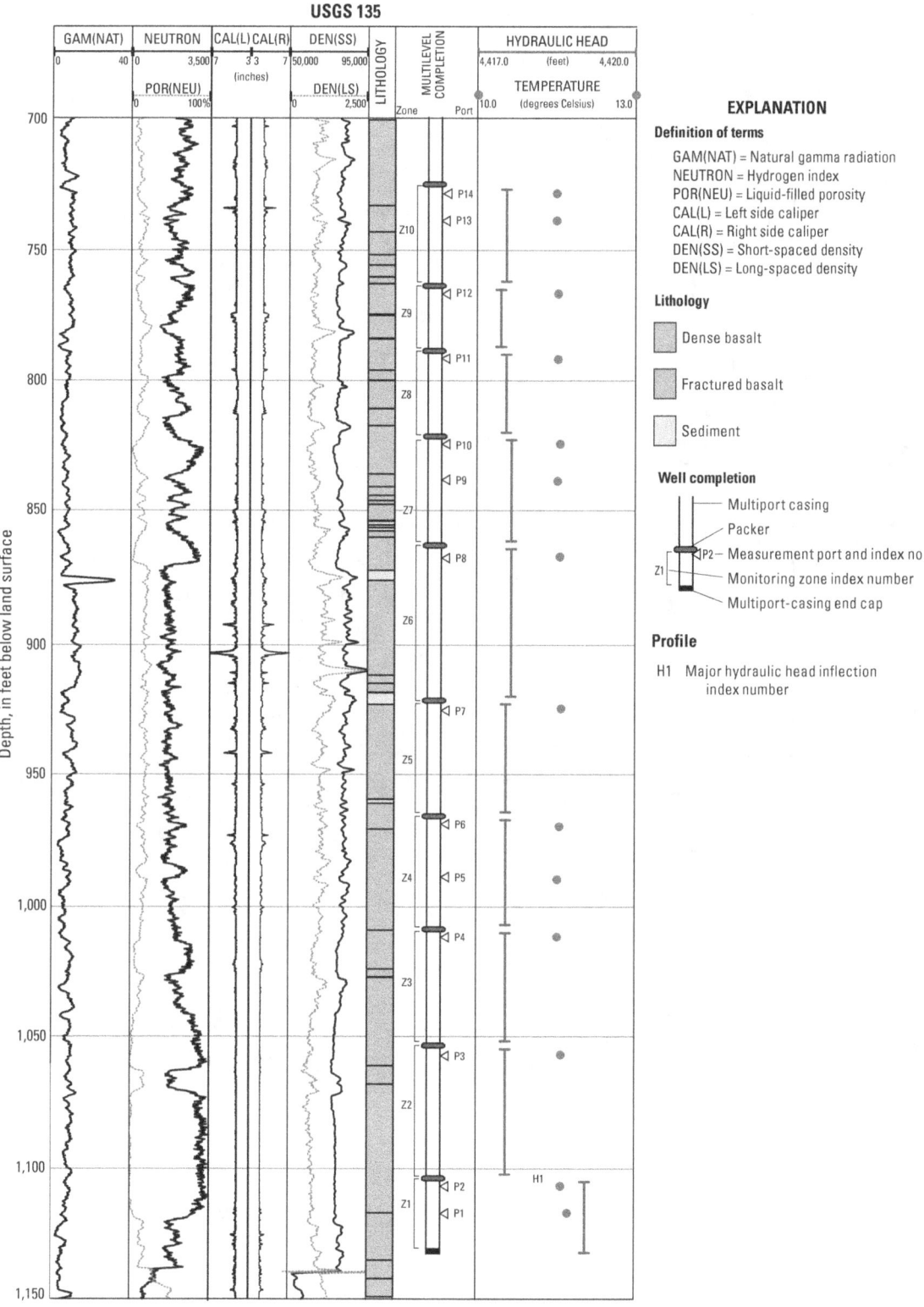

Figure 14. Geophysical traces of natural gamma, neutron, caliper, and gamma-gamma dual density; lithology log; multilevel completion; and hydraulic head and water temperature profiles for borehole USGS 135, Idaho National Laboratory, Idaho, June 2010.

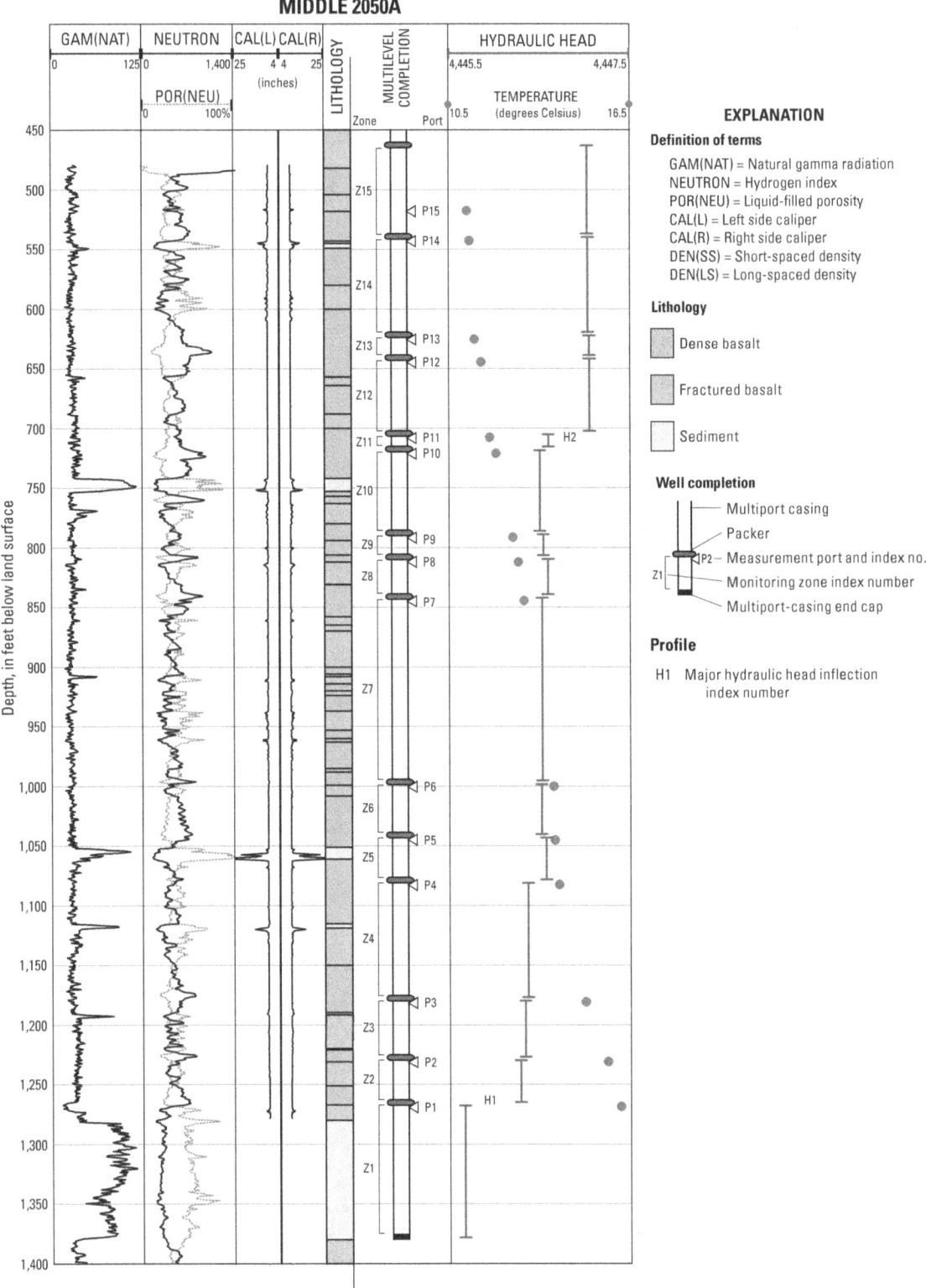

Figure 15.   Geophysical traces of natural gamma, neutron, caliper, and gamma-gamma dual density; lithology log; multilevel completion; and hydraulic head and water temperature profiles for borehole MIDDLE 2050A, Idaho National Laboratory, Idaho, June 2010.

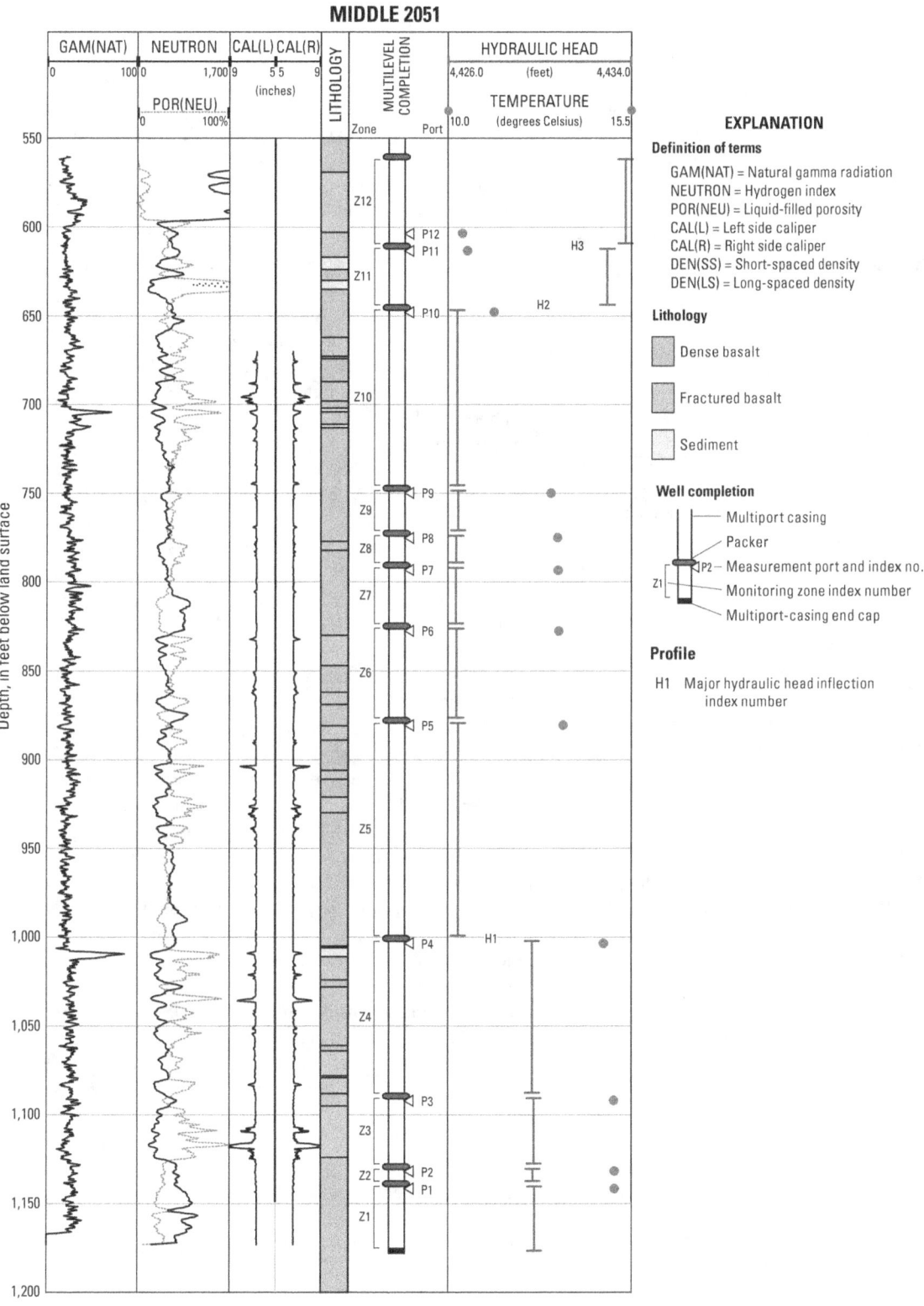

Figure 16.   Geophysical traces of natural gamma, neutron, caliper, and gamma-gamma dual density; lithology log; multilevel completion; and hydraulic head and water temperature profiles for borehole MIDDLE 2051, Idaho National Laboratory, Idaho, June 2010.

Borehole Profile Comparison

Hydraulic head and temperature profiles were compared between boreholes to better understand the groundwater flow system near the INL. The lithologic logs and profiles for all MLMS boreholes are shown in figure 17. The methods used to evaluate head profile comparisons were first described in Fisher and Twining (2011). Head values for the June 2010 measurement period (September 2010 for borehole USGS 108) were normalized to the mean head value for each zone. The mean-shifted hydraulic head of a profile, $\hat{H}$, is expressed as:

$$\hat{H}_i = H_i - \bar{H} \qquad \text{for } i = 1, \ldots, m \qquad (5)$$

where

H_i is the head measurement at port i in ft,

m is the total number of port head measurements in the MLMS, and

$\bar{H}$ is the mean head value in the profile (table 6) in ft, and defined as :

$$\bar{H} = \frac{1}{m} \sum_{i=1}^{m} H_i . \qquad (6)$$

The degree of vertical head change in a profile was evaluated using the range of its head measurements (table 6). The range in head for all MLMS boreholes was from 0.3 to 7.2 ft for boreholes USGS 132 and MIDDLE 2051, respectively. Previously reported head ranges were from 0.2 to 7.4 ft for boreholes USGS 103 and MIDDLE 2051, respectively (Fisher and Twining, 2011). The head range presented for June 2008 (Fisher and Twining, 2011) and the range presented in table 6 showed only slight variations, which is evidence that the profile shapes are relatively stable with time.

Lithologic units that appear to influence major head inflections typically coincided with sediment layers and dense basalt sequences that display minimal fracturing. For example, a 33-ft sediment layer in borehole USGS 133 and a 50-ft dense basalt layer in borehole USGS 135 account for about 99 percent of the total head change. Without knowing the true areal extent and transmissivity of lithologic units, it is difficult to determine where head changes might occur in a borehole. The determination is especially difficult in the ESRP aquifer because of its high level of heterogeneity, which is indicated by the large variability in profile shapes among boreholes.

The vertical connectivity between fracture networks of adjacent units is controlled by the geologic setting. Groundwater can move through vertical joints in dense basalt or through gaps in sediment layers that are not evident in core samples or geophysical data; however, MLMS data can detect slight variations in pressure. Low vertical gradients generally indicate potential vertical connectivity and flow,

and large gradient inflections indicate zones of relatively lower vertical connectivity, where vertical flow is potentially retarded (Fisher and Twining, 2011). The MLMS data reveal that zones that primarily are composed of fractured basalt displayed relatively small vertical head differences and flow was dominantly horizontal. For example, in borehole USGS 135, the head range above the H1 inflection was 0.1 (fig. 14). Other MLMS boreholes that displayed relatively small vertical head differences include USGS 103 (0.6 ft), USGS 105 (0.4 ft), and USGS 132 (0.3 ft) (table 6). Relatively moderate head ranges were reported in boreholes USGS 108 (0.9 ft), USGS 134 (0.9 ft), and MIDDLE 2050A (1.1 ft). In borehole USGS 108, inflections H1 and H2 occur at or near sediment layers; however, inflection H3 occurs where steel casing is extended to a depth of 760 ft bls, so borehole construction may have contributed to localized change in head (fig. 10). In borehole USGS 134, head values gradually decreased with depth above the H2 inflection and remained relatively constant below; whereas, in borehole MIDDLE 2050A, head values incrementally decreased with depth and remained relatively constant between large sediment layers (figs. 13 and 15). The highest head ranges were in boreholes USGS 133 (6.2 ft) and MIDDLE 2051 (7.2 ft) (figs. 12 and 16). The high head ranges for these boreholes are attributed to poor vertical connectivity between fracture networks, the result of sediment layering restricting vertical flow.

Mean groundwater temperature for the nine MLMS boreholes ranged from 11.5 to 13.8°C for boreholes USGS 135 and USGS 134, respectively (table 6). Vertical connectivity between adjacent aquifer units controlled the degree of convective or conductive heat transfer. In an idealized aquifer-confining layer system, where groundwater is mostly confined in aquifers and flow is dominantly horizontal, the temperature profile is dominated by upward conductive heat transfer and distributed linearly. When confining layers are not perfectly impervious, however, the temperature profile will be affected by convective heat transfer resulting from the vertical movement of groundwater (Bredehoeft and Papadopulos, 1965; Ge, 1998).

Temperature profiles in boreholes USGS 133, USGS 134, MIDDLE 2050A, and MIDDLE 2051 mostly follow a linear conductive trend with temperatures increasing with depth (fig. 17). These trends were similar to those reported in Fisher and Twining (2011), suggesting temperature in the ESRP aquifer remains constant with time near these locations. In borehole MIDDLE 2051, the limited vertical connectivity between adjacent aquifer units produces a strong linear conductive trend; however, within each aquifer unit, convective heat transfer dominates and temperatures remain relatively stable. The proximity of borehole USGS 134 to the northwest mountain front (fig. 1), where the saturated aquifer thickness thins and groundwater velocities are relatively small, results in an overall temperature that is approximately 1.5°C higher than the average groundwater temperature of the ESRP, medium aquifer temperature from 127 monitoring wells on the INL was 12.7 °C (Davis, 2010).

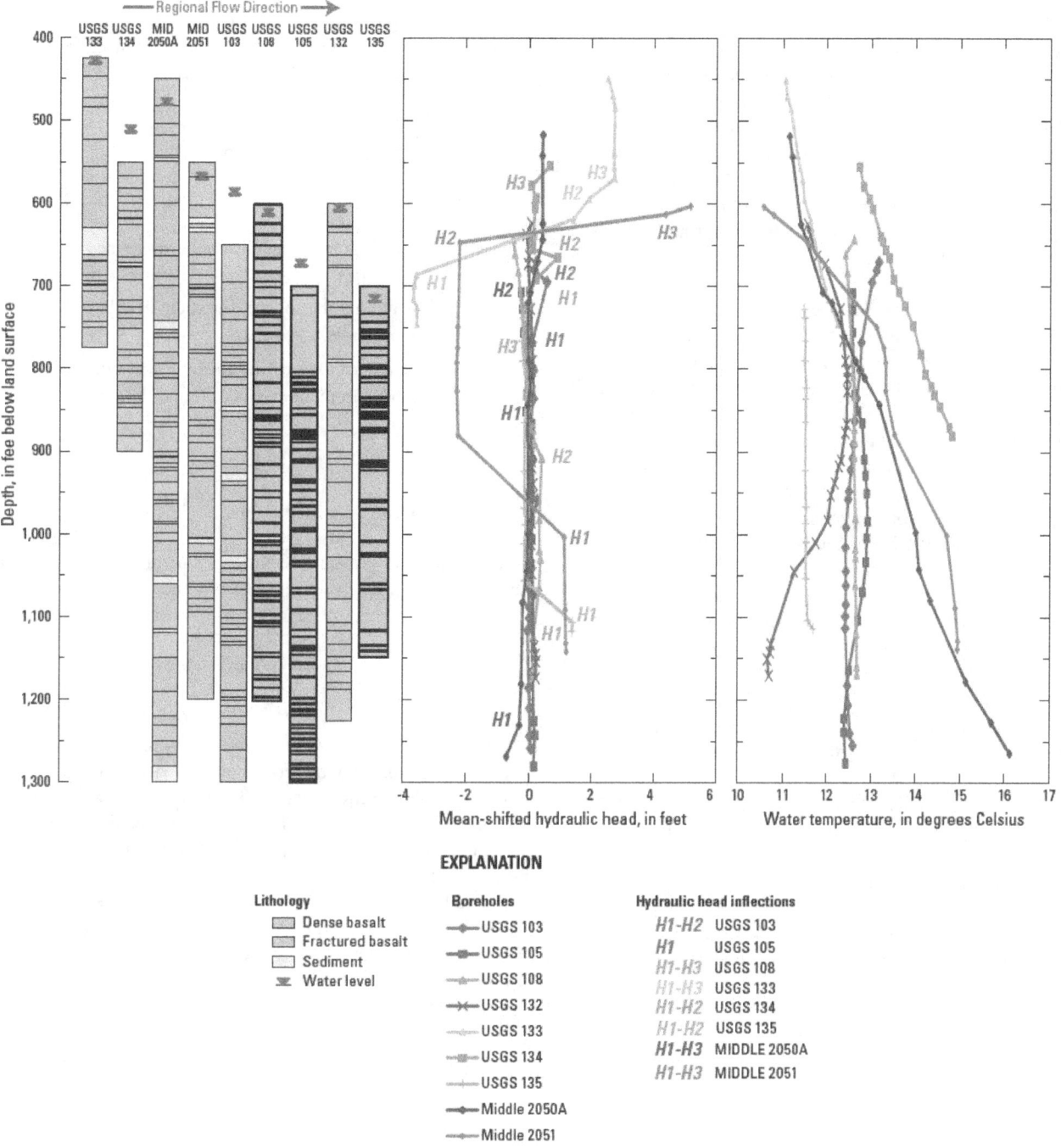

Figure 17.   Comparison among borehole lithology logs and mean-shifted hydraulic head and water temperature profiles for boreholes USGS 103, USGS 105, USGS 108, USGS 132, USGS 133, USGS 134, USGS 135, MIDDLE 2050A, and MIDDLE 2051, Idaho National Laboratory, Idaho, June 2010.

In boreholes USGS 103, USGS 105, USGS 108, USGS 132, and USGS 135, temperature profiles were mostly dominated by convective heat transfer, where fluid flow through the fractured media significantly altered the geothermal field (fig. 7). All five boreholes were completed within an area identified as the axial volcanic high (fig.1), a region characteristic of high volcanic activity and minimal sediment deposition. Lithologic data for all boreholes reveal sediment content of 5 percent or less and MLMS hydraulic head data suggest good connectivity between fracture networks of adjacent units.

Groundwater temperatures in boreholes USGS 132 and USGS 105 show a reversal in the convective heat transfer gradient in the lower half of the borehole profile; whereas, a similar reversal (opposite direction) was observed in the upper half of borehole USGS 103 (fig. 7). These temperature profiles suggest a stratified groundwater component and mixing of water types along the southern boundary of the INL (fig. 1, fig. 7). The decrease in temperature with depth in boreholes USGS 132 and USGS 105 is attributed to the location of the boreholes in a transition area; where the aquifer thickness (table 6) and transmissivity rapidly increase, flow has a downward component, and low-velocity tributary water mixes with high-velocity regional groundwater (Fisher and Twining, 2011). The vertical head and temperature characteristics of the regional groundwater are best reflected in the profiles of borehole USGS 103, the farthest MLMS borehole from the tributary valleys along the northwestern edge of the ESRP. The groundwater in borehole USGS 108 reflects an isothermal temperature profile that is well mixed, as it only changes 0.3°C in about 500 ft.

The mean groundwater temperature in borehole USGS 135 is 11.5°C, and shows a range of 0.2°C over about 400 ft. The groundwater temperature reflects convective heat transfer, the result of well-connected fracture networks for most of the borehole. The low average temperature and range suggest borehole USGS 135 is affected by groundwater originating from tributary valley inputs to the north (fig. 1).

The temporal correlation among MLMSs was analyzed for head data collected from 2007 to 2010 using the normalized mean hydraulic head, $\bar{H}$; these values of head were calculated for each MLMS and are expressed as:

$$\hat{\bar{H}}_t = \frac{\bar{H}_t - \bar{\bar{H}}}{s} \qquad \text{for } t = 1,\dots,n \qquad (7)$$

where

$\bar{H}_t$ is the profiles mean head for measurement event t in ft (eq. 6),

n is the total number of measurement events at the borehole,

$\bar{\bar{H}}$ is the mean of the profiles mean head values for all measurement events in ft and defined as

$$\bar{\bar{H}} = \frac{1}{n}\sum_{t=1}^{n}\bar{H}_t \; ; \qquad (8)$$

And s is the standard deviation of the profiles mean head values for all measurement events in ft and defined as

$$s = \sqrt{\frac{1}{n}\sum_{t=1}^{n}\left(H_t - \bar{\bar{H}}\right)} . \qquad (9)$$

The values of $\hat{\bar{H}}$ for each of the MLMS boreholes (fig. 18; appendix G) suggest a moderate positive correlation among all boreholes, which reflects regional fluctuations in water levels in response to seasonality. The temporal correlation is stronger when the location of boreholes in the aquifer is considered. For example, boreholes USGS 103, USGS 105, and USGS 108, and USGS 132 are in an area of highly permeable media with a relatively large saturated thickness, and a relatively rapid groundwater flow. As expected, the correlation between these four wells is strongly positive with a temporal trend that differs slightly from the other boreholes.

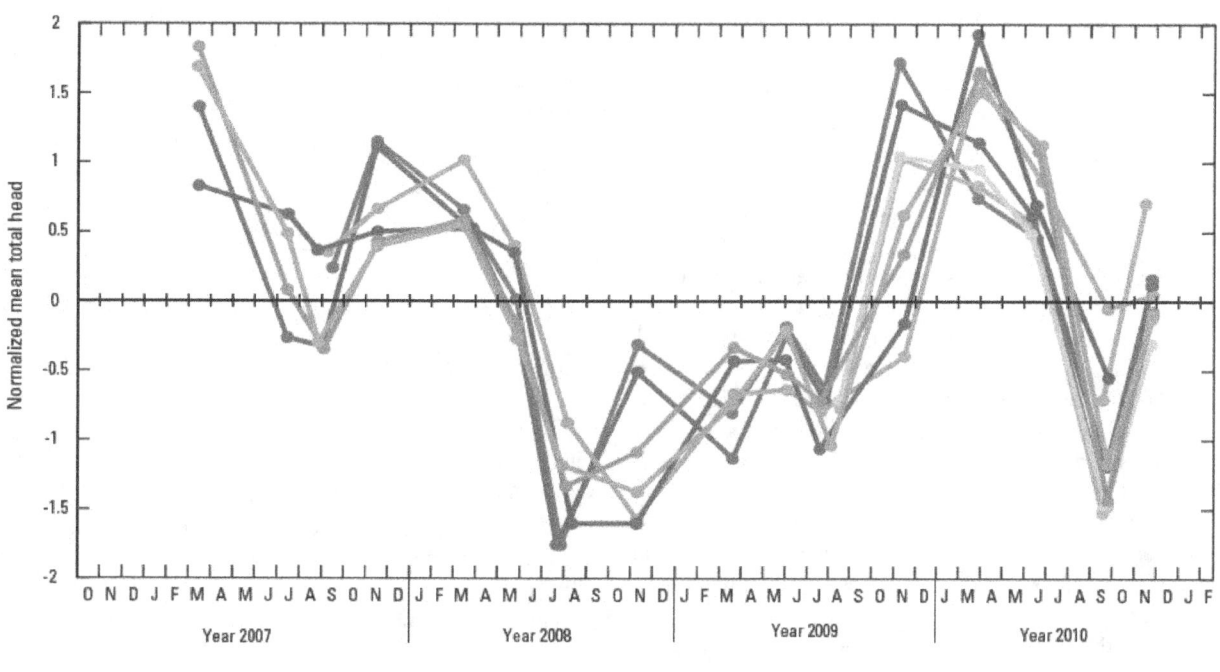

Figure 18.   Quarterly values of the normalized mean hydraulic head at boreholes USGS 103, USGS 105, USGS 108, USGS 132, USGS 133, USGS 134, USGS 135, MIDDLE 2050A, and MIDDLE 2051, Idaho National Laboratory, Idaho, 2008–10.

Summary

During 2009–10, quarterly depth-discrete measurements of hydraulic head (head) and water temperature were collected from multilevel monitoring systems (MLMSs) installed in nine boreholes at the Idaho National Laboratory (INL) in cooperation with the U.S. Department of Energy. The cored boreholes are located in the fractured basalts and interbedded sediments of the eastern Snake River Plain aquifer and were completed to depths ranging from 818 to 1,427 ft below land surface. Head and temperature measurements were recorded in 120 hydraulically isolated monitoring zones located 448.0 to 1,377.6 ft below land surface. The MLMS provides an approach for measurement of heads in the aquifer system that is needed as part of the characterization of the areal extent of contaminant plumes at the INL. Completion depths for MLMS boreholes exceed those of the average INL monitoring wells; therefore, the additional information on deep flow and contaminant transport conditions will assist ongoing efforts to better characterize the potential for downward plume migration and the extent of contamination in the aquifer.

Quarterly head and temperature profiles reveal unique vertical patterns for examination of the aquifer's complex basalt and sediment stratigraphy, proximity to aquifer recharge and discharge, and groundwater flow; these features contribute to some of the localized variability even though the general profile shape remained consistent over the measured time frame. Major inflections in the head profiles almost always coincided with low-permeability sediment layers and occasionally with thick sequences of dense basalt. However, the presence of a sediment layer or dense basalt layer was insufficient for identifying the location of a major head change within a borehole without knowing the true areal extent and relative transmissivity of the lithologic unit. Temperature profiles for boreholes completed within the Big Lost Trough indicate linear conductive trends; whereas, temperature profiles for boreholes completed within the axial volcanic high indicate mostly convective heat transfer resulting from the vertical movement of groundwater. Additionally, temperature profiles provide evidence for stratification and mixing of water types along the southern boundary of the Idaho National Laboratory.

The amount of vertical head and temperature change were examined for each of the nine multilevel monitoring systems. The vertical head gradients were defined for the major inflections in the head profiles and were as high as 2.1 feet per foot Groundwater temperatures in all boreholes ranged from 10.2 to 16.3°C. Low vertical head gradients indicate potential vertical connectivity and flow, and large gradient inflections indicated zones of relatively lower vertical connectivity. Generally, zones that primarily are composed of fractured basalt displayed relatively small vertical head differences, inferring flow is dominantly horizontal. Large head differences were attributed to poor vertical connectivity between fracture units which are the result of sediment layering and/or dense basalt.

Normalized mean hydraulic head values were analyzed for all nine multilevel monitoring wells for the period of record (2007–10). The mean head values suggest a moderately positive correlation among all boreholes, which reflects regional fluctuations in water levels in response to seasonal climatic changes. However, the temporal trend is slightly different depending on the well location; wells located along the southern boundary within the axial volcanic high show a strongly positive correlation.

Acknowledgments

The authors gratefully acknowledge Matt Gilbert and Jayson Blom of the USGS INL Project Office for their help with the installation of the MLMSs, and Jayson Blom, Amy Wehnke, Neil Maimer, and Betty Tucker, also of the USGS INL Project Office, for data collection. The authors also would like to acknowledge Schlumberger Water Services for their technical assistance with the Westbay™ Systems.

References Cited

Ackerman, D.J., 1991, Transmissivity of the Snake River Plain aquifer at the Idaho National Engineering Laboratory, Idaho: U.S. Geological Survey Water-Resources Investigations Report 91–4058 (DOE/ID22097), 35 p. (Also available at http://pubs.er.usgs.gov/publication/wri914058.)

Anderson, S.R., Ackerman, D.J., Liszewski, M.J., and Freiburger, R.M., 1996, Stratigraphic data for wells at and near the Idaho National Engineering Laboratory, Idaho: U.S. Geological Survey Open-File Report 96–248 (DOE/ID–22127), 27 p. and 1 diskette. (Also available at http://pubs.er.usgs.gov/publication/ofr96248.)

Anderson, S.R., and Liszewski, M.J., 1997, Stratigraphy of the unsaturated zone and the Snake River Plain Aquifer at and near the Idaho National Engineering Laboratory, Idaho: U.S. Geological Survey Water-Resources Investigations Report 97-4183 (DOE/ID-22142), 65 p. (Also available at http://pubs.er.usgs.gov/publication/wri974183.)

Anderson, S.R., Kuntz, M.A., and Davis, L.C., 1999, Geologic controls of hydraulic conductivity in the Snake River Plain aquifer at and near the Idaho National Engineering and Environmental Laboratory, Idaho: U.S. Geological Survey Water-Resources Investigations Report 99–4033 (DOE/ID–22155), 38 p. (Also available at http://pubs.er.usgs.gov/publication/wri994033.)

Bartholomay, R.C., Tucker, B.J., Ackerman, D.J., and Liszewski, M.J., 1997, Hydrologic conditions and distribution of selected radiochemical and chemical constituents in water, Snake River Plain aquifer, Idaho National Engineering Laboratory, Idaho, 1992 through 1995: U.S. Geological Survey Water-Resources Investigations Report 97-4086 (DOE/ID-22137), 57 p. (Also available at http://pubs.er.usgs.gov/publication/wri974086.)

Bartholomay, R.C., Tucker, B.J., Davis, L.C., and Greene, M.R., 2000, Hydrologic conditions and distribution of selected constituents in water, Snake River Plain aquifer, Idaho National Engineering and Environmental Laboratory, Idaho, 1996 through 1998: U.S. Geological Survey Water-Resources Investigations Report 2000–4192 (DOE/ID–22167), 52 p. (Also available at http://pubs.er.usgs.gov/publication/wri004192.)

Bartholomay, R.C., and Twining, B.V., 2010, Chemical constituents in groundwater from multiple zones in the eastern Snake River Plain aquifer at the Idaho National Laboratory, Idaho, 2005–08: U.S. Geological Survey Scientific Investigations Report 2010–5116, 82 p. (Also available at http://pubs.er.usgs.gov/publication/sir20105116.)

Bredehoeft, J.D., and Papadopulos, I.S., 1965, Rates of vertical groundwater movement estimated from the earth's thermal profile: Water Resources Research, v. 1, p. 325–328.

Busenberg, Eurybiades, Plummer, L.N., and Bartholomay, R.C., 2001, Estimated age and source of the young fraction of ground water at the Idaho National Engineering and Environmental Laboratory: U.S. Geological Survey Water-Resources Investigations Report 2001–4265 (DOE/ID–22177), 144 p. (Also available at http://pubs.er.usgs.gov/publication/wri014265.)

Cecil, L.D., Frape, S.K., Drimmie, R., Flatt, H., and Tucker, B.J., 1998, Evaluation of archived water samples using chlorine isotopic data, Idaho National Engineering and Environmental Laboratory, Idaho, 1966–93: U.S. Geological Survey Water-Resources Investigations Report 98–4008 (DOE/ID–22147), 27 p. (Also available at http://pubs.er.usgs.gov/publication/wri984008.)

Cecil, L.D., Welhan, J.A., Green, J.R., Frape, S.K., and Sudicky, E.R., 2000, Use of chlorine-36 to determine regional-scale aquifer dispersivity, eastern Snake River Plain aquifer, Idaho/USA: Nuclear Instruments and Methods in Physics Research, section B, v. 172, issue 1–4, p. 679–687.

Davis, L.C., 2008, An update of hydrologic conditions and distribution of selected constituents in water, Snake River Plain aquifer and perched–water zones, Idaho National Laboratory, Idaho, emphasis 2002–05: U.S. Geological Survey Scientific Investigations Report 2008–5089 (DOE/ID–22203), 7 p. (Also available at http://pubs.er.usgs.gov/publication/sir20085089.)

Davis, L.C., 2010, An update of hydrologic conditions and distribution of selected constituents in water, Snake River Plain aquifer and perched groundwater zones, Idaho National Laboratory, Idaho, emphasis 2006–08: U.S. Geological Survey Scientific Investigations Report 2010–5197 (DOE/ID–22212), 80 p. (Also available at .)

Duke, C.L., Roback, R.C., Reimus, P.W., Bowman, R.S., McLing, T.L., Baker, K.E., and Hull, L.C., 2007, Elucidation of flow and transport processes in a variably saturated system of interlayered sediment and fractured rock using tracer tests: Vadose Zone Journal, v. 6, no. 4, p. 855–867.

Fisher, J.C., and Twining, B.V., 2011, Multilevel groundwater monitoring of hydraulic head and temperature in the eastern Snake River Plain aquifer, Idaho National Laboratory, Idaho, 2007–08: U.S. Geological Survey Scientific Investigations Report 2010–5253, 62, p. (Also available at http://pubs.er.usgs.gov/publication/sir20105253.)

Garabedian, S.P., 1986, Application of a parameter–estimation technique to modeling the regional aquifer underlying the eastern Snake River Plain, Idaho: U.S. Geological Survey Water-Supply Paper 2278, 60 p. (Also available at http://pubs.er.usgs.gov/publication/wsp2278.)

Ge, Shemin, 1998, Estimation of groundwater velocity in localized fracture zones from well temperature profiles: Journal of Volcanology and Geothermal Research, v. 84, p. 93-101.

Geslin, J.K., Link, P.K., Riesterer, J.W., Kuntz, M.A., and Fanning, C.M., 2002, Pliocene and Quaternary stratigraphic architecture and drainage systems of the Big Lost Trough, northeastern Snake River Plain, Idaho, in Link, P.K., and Mink, L.L., eds., Geology, hydrogeology, and environmental remediation—Idaho National Engineering and Environmental Laboratory, Eastern Snake River Plain, Idaho: Boulder, Colo., Geological Society of America Special Paper 353, p. 11–26.

Gianniny, G.L., Geslin, J.K., Riesterer, J.W., Link, P.K., and Thackray, G.D., 1997, Quaternary surficial sediments near Test Area North (TAN) northeastern Snake River Plain—An actualistic guide to aquifer characterization, in Sharma, S., and Hardcastle, J.H., eds., Symposium on engineering geology and geotechnical engineering, 32d, Boise, Idaho, March 26–28, 1997, Proceedings: Pocatello, Idaho, Idaho State University College of Engineering, v. 44, p. 29.

Hughes, S.S., Smith, R.P., Hackett, W.R., and Anderson, S.R., 1999, Mafic volcanism and environmental geology of the eastern Snake River Plain, Idaho, *in* Hughes S.S., and Thackray, G.D., eds., Guidebook to the geology of eastern Idaho: Pocatello, Idaho, Idaho Museum of Natural History, p. 143–168.

Hughes, S.H., Wetmore, P.H., and Casper, J.L., 2002, Evolution of Quaternary tholeiitic basalt eruptive centers on the eastern Snake River Plain, Idaho, *in* Bonnichsen, B., White, C.M., and McCurry, M., eds., Tectonic and magmatic evolution of the Snake River Plain volcanic province: Idaho Geological Survey Bulletin, v.30, p. 363–385.

Knutson, C.F., McCormick, K.A., Crocker, J.C., Glenn, M.A., and Fishel, M.L., 1992, 3D RWMC vadose zone modeling (including FY–89 to FY–90 basalt characterization results): E G and G Idaho, Inc., Report EGG–ERD–10246 [variously paged].

Kuntz, M.A., Skipp, Betty, Lanphere, M.A., Scott, W.E., Pierce, K.L., Dalrymple, G.B., Champion, D.E., Embree, G.F., Page, W.R., Morgan, L.A., Smith, R.P., Hackett, W.R., and Rodger, D.W., 1994, Geologic map of the Idaho National Engineering Laboratory and adjoining areas, eastern Idaho: U.S. Geological Survey Miscellaneous Investigations Map I-2330, scale 1:100,000.

Mann, L.J., and Beasley, T.M., 1994, Iodine–129 in the Snake River Plain aquifer at and near the Idaho National Engineering Laboratory, Idaho, 1990–91: U.S. Geological Survey Water-Resources Investigations Report 94–4053 (DOE/ID–22115), 27 p. (Also available at http://pubs. er.usgs.gov/publication/wri944053.)

McCutcheon, S.C., Martin, J.L., and Barnwell, T.O., Jr., 1993, Water quality, *in* Maidment, D.R., ed., Handbook of hydrology: New York, McGraw–Hill, chap. 11, p. 3–73.

Nimmo, J.R., Perkins, K.S., Rose, P.E., Rousseau, J.P., Orr, B.R., Twining, B.V., and Anderson, S.R., 2002, Kilometer–scale rapid transport of naphthalene sulfonate tracer in the unsaturated zone at the Idaho National Engineering and Environmental Laboratory: Vadose Zone Journal, v. 1, no. 1, p. 89–101.

R Development Core Team, 2010, R—A language and environment for statistical computing: Vienna, Austria, R Foundation for Statistical Computing, ISBN 3–900051v07–0, accessed June 15, 2012, at http://www.R-project.org.

Robertson, J.B., Schoen, Robert, and Barraclough, J.T., 1974, The influence of liquid waste disposal on the geochemistry of water at the National Reactor Testing Station, Idaho, 1952–1970: U.S. Geological Survey Open-File Report 73–238 (IDO–22053), 231 p. (Also available at http://pubs. er.usgs.gov/publication/ofr73238.)

Self, Stephen, Keszthelyi, L., and Thordarson, Th., 1998, The importance of Pahoehoe: Annual Review of Earth Planetary Sciences, v. 26, p. 81–110.

U.S. Geological Survey, 1985, National water summary, 1984—Hydrologic events, selected water–quality trends, and ground–water resources: U.S Geological Survey Water-Supply Paper 2275, 467 p. (Also available at http://pubs. er.usgs.gov/publication/wsp2275.)

U.S. Geological Survey, 2012, Idaho National Laboratory Project Office Publications: Website, accessed September 20, 2012, at http://id.water.usgs.gov/projects/ INL/pubs.html.

Welhan, J.A., Farabaugh, R.L., Merrick, M.J., and Anderson, S.R., 2007, Geostatistical modeling of sediment abundance in a heterogeneous basalt aquifer at the Idaho National Laboratory, Idaho: U.S. Geological Survey Scientific Investigations Report 2006–5316 (DOE/ID–22201), 32 p. (Also available at http://pubs.er.usgs.gov/publication/ sir20065316.)

Whitehead, R.L., 1992, Geohydrologic framework of the Snake River Plain regional aquifer system, Idaho and eastern Oregon: U. S. Geological Survey Professional Paper 1408-B, 32 p. (Also available at http://pubs.er.usgs.gov/ publication/pp1408B.)

Appendixes

Appendix files are available for download at http://pubs.usgs.gov/sir/2012/5259.

Appendix A. Data Used to Calculate Pressure Probe Transducer Depths at Measurement Port Couplings, Boreholes USGS 105, USGS 108, and USGS 135, Idaho National Laboratory, Idaho, 2009–10

Appendix B. Field sheet used for data collection at multilevel monitoring boreholes, Idaho National Laboratory, Idaho

Appendix C. Calibration results for fluid pressure sensor, a component of the sampling probe used in boreholes USGS 103, USGS 105, USGS 108, USGS 132, USGS 133, USGS 134, USGS 135, MIDDLE 2050A, and MIDDLE 2051, Idaho National Laboratory, Idaho, 2008–11

Appendix D. Barometric pressure, water temperature, fluid pressure, and hydraulic head data from port measurements for boreholes USGS 103, USGS 105, USGS 108, USGS 132, USGS 133, USGS 134, USGS 135, MIDDLE 2050A, and MIDDLE 2051, Idaho National Laboratory, Idaho, 2009–10

Appendix E. Lithology Logs for Multilevel Groundwater Monitoring Boreholes USGS 103, USGS 105, USGS 108, USGS 132, USGS 133, USGS 134, USGS 135, MIDDLE 2050A, and MIDDLE 2051, Idaho National Laboratory, Idaho, 2007–08

Appendix F. Vertical hydraulic head gradient data between adjacent monitoring zones for boreholes USGS 103, USGS 105, USGS 108, USGS 132, USGS 133, USGS 134, USGS 135, MIDDLE 2050A, and MIDDLE 2051, Idaho National Laboratory, Idaho, June 2010 and September 2010 for USGS 108

Appendix G. Quarterly mean and normalized mean hydraulic head values for boreholes USGS 103, USGS 105, USGS 108, USGS 132, USGS 133, USGS 134, USGS 135, MIDDLE 2050A, and MIDDLE 2051, Idaho National Laboratory, Idaho, 2007–10